$6.75

BORON-NITROGEN COMPOUNDS

ANORGANISCHE UND ALLGEMEINE CHEMIE
IN EINZELDARSTELLUNGEN
HERAUSGEGEBEN VON
MARGOT BECKE-GOEHRING
BAND VI

BORON-NITROGEN COMPOUNDS

BY

Dr. KURT NIEDENZU and Dr. JOHN W. DAWSON

U.S. ARMY RESEARCH OFFICE, DUKE UNIVERSITY
DURHAM, NORTH CAROLINA

1965

NEW YORK

ACADEMIC PRESS INC., PUBLISHERS

BERLIN · HEIDELBERG · NEW YORK
SPRINGER-VERLAG

SPRINGER-VERLAG

BERLIN · HEIDELBERG · NEW YORK

Published in the U.S.A. and Canada by

ACADEMIC PRESS INC., PUBLISHERS

111 Fifth Avenue, New York, New York 10003

© BY SPRINGER-VERLAG

BERLIN · HEIDELBERG

1965

Printed in Germany

Vorbemerkung des Herausgebers

Die Chemie der Bor-Stickstoff-Verbindungen geht in ihren präparativen Anfängen auf ALFRED STOCK zurück und ihre Sonderstellung im Rahmen der anorganischen Chemie wurde zuerst von EGON WIBERG erkannt. Die Bor-Stickstoff-Verbindungen sind in der Folgezeit von einer ganzen Anzahl hervorragender Chemiker bearbeitet worden und seit etwa acht Jahren kann man von einer fast stürmischen Entwicklung auf diesem Gebiet sprechen. Obgleich heute die präparativen Möglichkeiten noch lange nicht erschöpft sind und obgleich man immer noch recht wenig über die Bindungsverhältnisse, die die Besonderheiten der Bor-Stickstoff-Bindung bestimmen, weiß, erschien es doch zweckmäßig, eine zusammenfassende Übersicht über dieses Gebiet der anorganischen Chemie zu geben. Es ist zu hoffen, daß das vorliegende Bändchen alle diejenigen, die sich für das weite Feld der Chemie der nichtmetallischen Elemente interessieren, über den Stand unserer Kenntnis auf dem Gebiet der Bor-Stickstoff-Verbindungen informieren und sie zu neuen Versuchen und Überlegungen anregen möge.

November 1964

M. BECKE-GOEHRING

Preface

Although the chemistry of boron is still relatively young, it is developing at a pace where even specific areas of research are difficult to compile into a monograph. Besides the boron hydrides, boron-nitrogen compounds are among the most fascinating derivatives of boron. Nitrogen compounds exist in a wide variety of molecular structures and display many interesting properties. The combination of nitrogen and boron, however, has some unusual features that are hard to match in any other combination of elements. This situation was first recognized by ALFRED STOCK and it seems proper to pay tribute to his outstanding work in the area of boron chemistry. One should realize that about forty years ago, STOCK and his coworkers had to develop completely new experimental techniques and that no guidance for the interpretation of their rather unusual data had been advanced by theoretical chemists.

In this monograph an attempt has been made to explore the general characteristics of structure and the principles involved in the preparation and reactions of boron-nitrogen compounds. It was a somewhat difficult task to select that information which appears to be of the most interest to "inorganic and general chemistry" since the electronic relationship between a boron-nitrogen and a carbon-carbon grouping is reflected in the "organic" character of many of the reactions and compounds. Due to this fascinating interplay and the limited scope of a monograph, the description of some areas of boron-nitrogen chemistry has been restricted to a very brief resume. Also, the chemistry of the nitrogen derivatives of the higher boron hydrides has been deliberately curtailed in the present description; it is felt that this area should more properly be considered as boron hydride chemistry rather than that of boron-nitrogen compounds.

This monograph is being offered as an example of what can evolve from a close integration of several areas of chemistry. In boron-nitrogen chemistry rarely are there examples that are strictly inorganic, organic or physical in nature. Rather, this area of chemistry examplifies what can be accomplished through healthy cross-fertilization. The influences of the various subdisciplines have contributed much toward promoting a lively interest in the chemistry of boron-nitrogen compounds.

The authors gratefully acknowledge the invitation by Prof. M. BECKE-GOEHRING for us to write this monograph and particularly for her meticulous and careful review of the manuscript and her helpful and constructive suggestions leading to its improvement.

August 1964

JOHN W. DAWSON
KURT NIEDENZU

Table of Contents

Table of Contents

Introduction

Boron, a black metallic appearing element, was prepared by GAY-LUSSAC and THENARD as far back as 1808[1]. However, its best known derivative, borax, was known to the Arabs as TINKAL and was used in melting processes in the sixteenth century. Boric acid was prepared by HOMBERG[2] in 1702 and found wide application as "sal sedativum". Also in 1808, DAVY[3] obtained amorphous boron by the electrolysis of boric acid and a crystalline modification was described by WÖHLER and SAINTE-CLAIRE DEVILLE in 1856[4]. Large deposits of boron minerals are found in various parts of the world. The most important stocks are located in Tuscany (Italy), in the southwestern part of the United States, and in Northern Chile and neighboring parts of Argentina, Peru and Bolivia. Smaller deposits of commercial interest are found in Turkey, Central Asia (Tibet) and near Stassfurt, Germany. The percentage of boron in the earth's crust is estimated at about 0.001%.

Despite the widespread occurrence of boron compounds and the early recognition of boron as an element, studies of boron chemistry have long been confined to a relatively small area. This was due mainly to the instability of many boron compounds towards hydrolysis or oxidation and the inherent difficulties in handling many of the materials. However, when STOCK developed his now famous experimental vacuum techniques, a door was opened for the investigation of a completely new field of boron chemistry. The boron hydrides were exhaustively studied and, it is indicative of STOCK's outstanding work that, until a few years ago, no new boron hydride had been discovered for about thirty years.

Reports on boron-nitrogen compounds date back about 150 years. However, it was the inspired work of ALFRED STOCK and his coworkers[5] and the experimental techniques developed in their basic investigation of boron hydrides some forty years ago which provided the necessary impetus for a detailed study of boron-nitrogen derivatives. In 1926 STOCK and POHLAND[6] studied the reaction of diborane with ammonia which resulted in the discovery of borazine, (—BH—NH—)$_3$, the "inorganic

---

[1] GAY-LUSSAC, J. L., and L. J. THENARD: Ann. Chim. Phys. [1] **68**, 169 (1808).

[2] HOMBERG, G.: Mem. Acad. polon Sci Letters. Cl. Sci. math. natur., Ser. A. **33**, 50 (1702).

[3] DAVY, H.: Philos. Trans. Roy. Soc. London Ser. A. **98**, 333 (1808).

[4] WÖHLER, F., and H. SAINTE-CLAIRE DEVILLE: C. R. hebd. Seances Acad. Sci. **43**, 1088 (1856).

[5] STOCK, A.: Hydrides of Boron and Silicon, Ithaca, N.Y.: Cornell University Press 1933.

[6] STOCK, A., and E. POHLAND: Ber. dtsch. chem. Ges. **59**, 2215 (1926).

benzene". This event might be considered as the birth of modern boron-nitrogen chemistry. Nevertheless, until about 1950, few research groups devoted any real effort to the exploration of this area. Since that time, however, an increasing number of scientists have become interested in boron-nitrogen compounds and recent methods have been developed which provide for at least partial replacement of the high vacuum techniques previously used in STOCK's classic work, by more conventional methods.

Historical Dates of Boron-Nitrogen Chemistry

	1809	GAY-LUSSAC	adduct of ammonia and trifluoro-borane
	1842	BALMAIN	first reports on boron nitride
	1850	WÖHLER, ROSE	characterisation of boron nitride
about	1905	STOCK, JOANNIS	studies on the interaction of ammonia with trihalogenoboranes
	1926	STOCK and POHLAND	synthesis of borazine
since	1935	WIBERG, SCHLESINGER	basic studies of boron-nitrogen compounds
	1955	BROWN and LAUBENGAYER	preparation of B-trichloroborazine with standard laboratory equipment
since	1956	DEWAR and coworkers	heteroaromatic boron-nitrogen compounds
	1958	PARRY and coworkers	structure of the diammoniate of diborane

It is of interest to note that the more recent efforts are independent of any commercial interest in boron-nitrogen compounds and are based primarily on the pursuit of basic research.

As direct neighbors of carbon in the Periodic Table of Elements, the combination of boron and nitrogen has the same number of electrons as the carbon-carbon entity. Moreover, the sum of the atomic radii of boron and nitrogen is of the same order of magnitude as that of two carbon atoms.

Elemental Characteristics of Boron, Carbon and Nitrogen

	number of outer electrons	covalent single-bond radii, Å	electronegativity
boron. . . .	3	0.88	2.0
carbon . . .	4	0.771	2.5
nitrogen . .	5	0.70	3.0

The tercovalent nitrogen has a pair of unshared electrons which are available to complete the octet of a tervalent, electron-deficient, boron atom. Thus a close relationship in behavior and characteristics between some substances having boron-nitrogen bonds and others with similar carbon-carbon linkages is to be expected since the two bonds are of similar size. Indeed, considerable attention has been drawn to the existence of a number of boron-nitrogen compounds similar in many respects to their analogous organic counterparts.

WIBERG[1] classified boron-nitrogen derivatives into the three major groups of amine-boranes (Borazane), aminoboranes (Borazene) and borazines (Borazole) according to the nature of the B—N bond in the molecules. *Amine-Boranes* are characterized by having a tercovalent boron bonded to a tercovalent nitrogen by the two unshared electrons of the nitrogen (I). This structure is roughly comparable to that of alkanes, in the sense that in alkanes two carbon atoms are linked by a single two-electron bond (II).

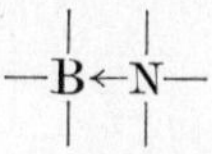

$$-\overset{|}{\underset{|}{B}}\!\leftarrow\!\overset{|}{\underset{|}{N}}- \qquad\qquad -\overset{|}{\underset{|}{C}}-\overset{|}{\underset{|}{C}}-$$

I 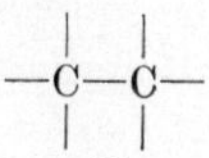   II

In the *Aminoborane* system, boron and nitrogen are linked together by one normal covalent bond; but the unshared electron pair of the nitrogen can participate in this linkage[2] thereby introducing a degree of double bond character (III). Aminoboranes, therefore, often exhibit physical characteristics analogous to those of alkenes (IV).

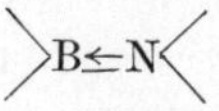

$$>\!\!B\!\Leftarrow\!\!N\!\!< \qquad\qquad >\!\!C\!=\!C\!\!<$$

III 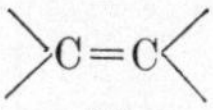   IV

The third major group of boron-nitrogen compounds comprises the cyclic *Borazines* (V), whose formal analogy to aromatic carbon compounds (VI) is illustrated by the structures.

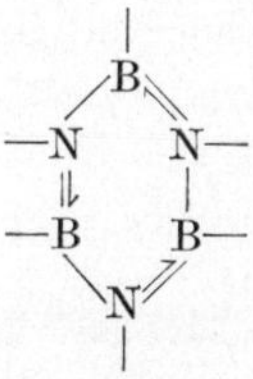

V 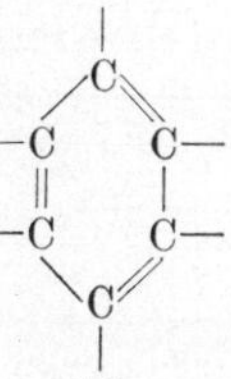   VI

However, any such comparison between boron-nitrogen compounds and organic materials should not be pursued excessively. Although B—N and C—C linkages are isoelectronic and very similar, they are not identical. The C—C bond involves atoms of the same element, while that of B—N resides between atoms of differing electronegativity and, therefore, the electron cloud representing the B—N bond is not symmetrical. Differences between the two systems might very well outweigh the similarities. Nevertheless, the concept of isosterism of B—N and C—C compounds serves well to effect a general classification of boron-nitrogen derivatives and has been adopted as a model for the following chapters.

It seems proper to mention in this introduction, two reactions which have been found to be of major importance in this area of chemistry.

---

[1] WIBERG, E.: Naturwissenschaften **35**, 182, 212 (1948).

[2] In the following, arrows (→) will be used to emphasize lone-pair electron bonding and to distinguish from the normal covalent bond illustrated by a straight line (—).

Boron compounds have a distinct tendency to disproportionate, apparently more so than those of any other nonmetallic element. This tendency is readily explained in terms of the electron deficiency of trivalent boron, which provides a center for nucleophilic attack. Although often annoying in preparative work, disproportionation reactions of boron compounds have nevertheless been put to good use. This is illustrated by the interaction of triorganoboranes with trihalogenoboranes to yield various organohalogenoboranes, which are not readily available by other synthetic methods. Of equal importance in boron-nitrogen chemistry is the transamination reaction. The potentialities of this reaction are only now beginning to be exploited. During the past three or four years, transamination has already provided for a remarkably simple access to several major classes of boron-nitrogen compounds.

Interest is also quickening in the extent to which certain boron-nitrogen compounds may be applied for technical purposes. This upsurge resulted mainly from the proposed military uses of boron compounds as high energy fuels. Although this application has not realized its full potential, diborane for example has now become commercially available. B-Trichloro-borazine is offered in developmental quantities and a variety of other boron-nitrogen compounds have appeared on the market. The use of boron nitride as a possible semiconductor is being explored and, in the search for inorganic and semi-inorganic polymers applicable for high temperature uses, various boron-nitrogen derivatives demonstrate interesting possibilities.

The reader interested in a more detailed study of the progress and aspects of boron chemistry is referred to some recent articles and books listed below.

Books:

GMELIN's Handbuch der anorganischen Chemie, "Bor", System Nummer 13, Ergänzungsband. Verlag Chemie: Weinheim 1954.
LIPSCOMB, W. N.: Boron Hydrides. New York/Amsterdam: W.A. Benjamin Inc. 1963.
BROWN, H. C.: Hydroboration. New York: W. A. Benjamin Inc. 1962.
GERRARD, W.: The Organic Chemistry of Boron. New York/London: Academic Press 1961.
Advances in Chemistry Series, Vol. 42, Boron-Nitrogen Chemistry. Washington, D.C.: American Chemical Society.
Advances in Chemistry Series, Vol. 32, Borax to Boranes. Washington, D.C.: American Chemical Society.

Articles:

MAITLIS, P. M.: Heterocyclic Organic Boron Compounds. Chem. Review **62**, 223 (1962).
GERRARD, W., and M. F. LAPPERT: Reactions of Boron Trichloride with Organic Compounds, Chem. Review. **58**, 1081 (1958).
STONE, F. G. A.: Stability Relationships amond Analogous Molecular Addition Compounds of Group III Elements, Chem. Review. **58**, 101 (1958).
WIBERG, E.: Neuere Entwicklungslinien der Borchemie. Experientia [Basel] Suppl. VII, 183 (1957).

Nomenclature of Boron-Nitrogen Compounds

In the following chapters, the nomenclature recommended by the Committee of the American Chemical Society on the Nomenclature of Organic Boron Compounds[1,2] will be used. The principles of this nomenclature, insofar as they concern boron-nitrogen compounds, are outlined below.

I. Linear Systems

1. The combining form of boron is "bor-" and is used to designate the presence of boron in a compound.

2. For such compounds which can formally be considered as substituted BH_3, the suffix "ane" is applied. Thus BCl_3 would be correctly named "trichloroborane", although in this case (and several others of common usage) the term "boron trichloride" is acceptable.

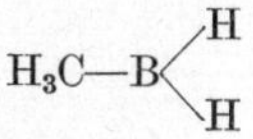

$$H_3C—B\begin{array}{c}H\\\\H\end{array} \qquad \text{methylborane}$$

$$(H_3C)_2N—B\begin{array}{c}N(CH_3)_2\\\\N(CH_3)_2\end{array} \qquad \text{tris(dimethylamino)borane}$$

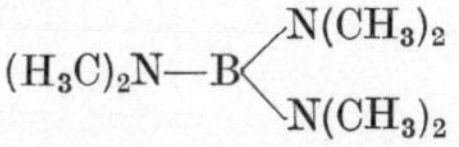

3. Whenever possible, substituents on the boron are named according to their accepted title as given in Chemical Abstracts. The order of the groups is governed by the practice used for substituted organic compounds.

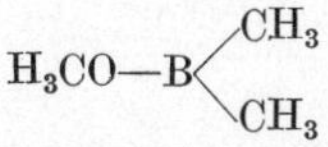

$$H_3CO—B\begin{array}{c}CH_3\\\\CH_3\end{array} \qquad \begin{array}{l}\text{methoxy-dimethylborane } or\\ \text{(methoxy)dimethylborane}\end{array}$$

$$\begin{array}{c}H_3C\\\\H_3C\end{array}N—B\begin{array}{c}CH_3\\\\CH_3\end{array} \qquad \text{(dimethylamino)dimethylborane}$$

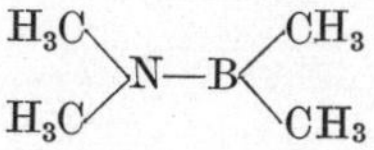

4. In addition compounds (i.e. donor-acceptor complexes), both molecules are given their proper names and the title of the addition compound is formed by indicating the donor molecule first, and then joining this to the name of the acceptor by a hyphen.

$(CH_3)_3N \cdot BF_3$ trimethylamine-trifluoroborane

$(CH_3)H_2N \cdot BH_2Cl$ methylamine-chloroborane

---

[1] Chem. Engng. News **32,** 1441 (1954).
[2] ibid. **34,** 560 (1956).

5. If a need exists to indicate the addition linkage, the simple hyphen is replaced by the symbols of the two elements forming the donor-acceptor bond; the symbols are joined by a hyphen and enclosed in parentheses.

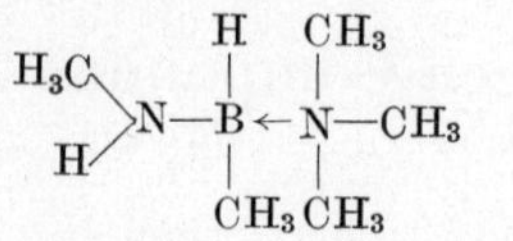

$$\begin{array}{c} H_3C \\ \diagdown \\ H \diagup \end{array} N\!-\!\overset{\displaystyle H}{\underset{\displaystyle CH_3}{B}}\!\leftarrow\!\overset{\displaystyle CH_3}{\underset{\displaystyle CH_3}{N}}\!-\!CH_3$$
   trimethylamine(N—B)-(methylamino)methylborane

6. Hydrogen functioning as a ligand is named "hydro-" and the name is placed last in the order of ligands.

$Na[BH_4]$ sodium hydroborate

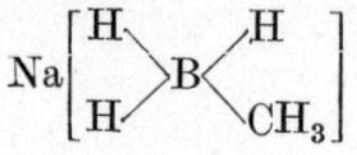

$$Na\left[\begin{array}{c} H \\ \diagdown \\ H \diagup \end{array} B \begin{array}{c} H \\ \diagdown \\ \diagup \\ CH_3 \end{array}\right]$$
   sodium methyltrihydroborate

II. Ring Systems

1. Names of small ring systems are most conveniently given by the practices established in the Ring Index.

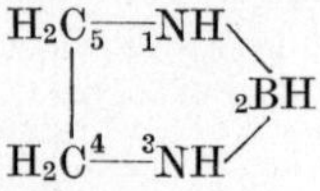

$$\begin{array}{c} H_2C_5 \overline{} {}_1NH \\ | \qquad \diagdown {}_2BH \\ H_2C_4 \!-\!{}_3NH \diagup \end{array}$$
   1,3,2-diazaborolidine *or*
   1,3,2-diazaboracyclopentane

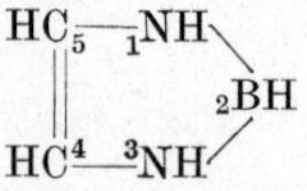

$$\begin{array}{c} HC_5 \overline{} {}_1NH \\ \| \qquad \diagdown {}_2BH \\ HC_4 \!-\!{}_3NH \diagup \end{array}$$
   $\varDelta$ 4-1,3,2-diazaboroline

2. In larger rings, especially when these are in the highest stage of hydrogenation, names based on those of the corresponding hydrocarbons may be used.

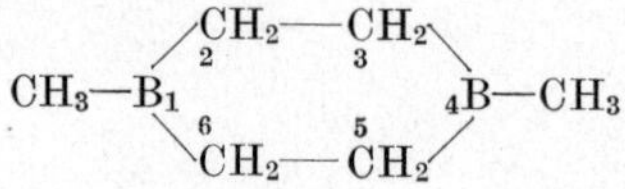

$$CH_3\!-\!B_1 \begin{array}{c} \diagup CH_2 \!-\!CH_2 \diagdown \\ {}_2 \qquad\qquad {}_3 \\ \\ {}_6 \qquad\qquad {}_5 \\ \diagdown CH_2 \!-\! CH_2 \diagup \end{array} {}_4B\!-\!CH_3$$
   1,4-dimethyl-1,4-diborinane *or*
   1,4-dimethyl-1,4-diboracyclohexane

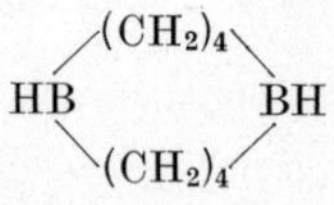

$$\begin{array}{c} \diagup (CH_2)_4 \diagdown \\ HB \qquad\qquad BH \\ \diagdown (CH_2)_4 \diagup \end{array}$$
   1,6-diboracyclodecane

3. Trivial names are to be used only if the ring possesses a particular feature (i.e. stability, great number of derivatives etc.).

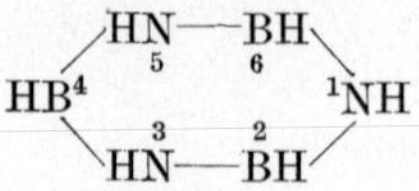

$$\begin{array}{c} \diagup HN\!-\!BH \diagdown \\ HB^4 \quad {}_5 \quad {}_6 \quad {}_1NH \\ \\ {}_3 \qquad {}_2 \\ \diagdown HN\!-\!BH \diagup \end{array}$$
   (hexahydro-*s*-triazatriborine)
   borazine (but not borazole)

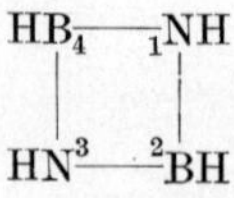

$$\begin{array}{c} HB_4 \overline{} {}_1NH \\ | \qquad | \\ HN^3 \!-\!{}_2BH \end{array}$$
   1,3-diaza-2,4-boretane *or*
   tetrahydro-*s*-diazadiborin

4. The numbering of the ring follows the usual practice for hetero-cyclic rings. In those cases where a high order of symmetry exists, a short form may be used.

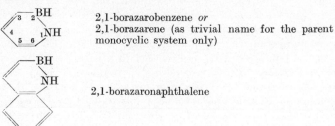

2,4,6-trichloroborazine *or*
B-trichloroborazine

5. When boron and nitrogen are incorporated in an aromatic six-membered ring system, nomenclature according to the Ring Index would not illustrate their essential features, i.e. that such compounds are isoelectronic with aromatic hydrocarbons. Consequently, the convention developed by DEWAR and DIETZ[1] is then adopted, wherein the compounds are named as derivatives of the isoelectronic hydrocarbon with the prefix "aro" to denote the existence of aromaticity in the system.

2,1-borazarobenzene *or*
2,1-borazarene (as trivial name for the parent monocyclic system only)

2,1-borazaronaphthalene

[1] DEWAR, M. J. S., and R. DIETZ: J. chem. Soc. (London) 1959, 2728.

Chapter I

Amine-Boranes and Related Structures

A. The Boron-Nitrogen Dative Bond

1. General Remarks

The ability of trivalent boron compounds of the type BR_3 to accept electron pairs from suitable electron donors has long been recognized. The driving force for such events is the tendency for boron to complete its outer shell of electrons in order to obtain the most favorable electronic configuration. Hence an extremely large number of molecular addition compounds of BR_3 molecules with a variety of nitrogen bases and, more general, nitrogen-containing compounds, have been described over the last 150 years. For instance, $H_3N \cdot BF_3$, the adduct of ammonia and trifluoroborane, was prepared by GAY-LUSSAC in 1809[1] and was studied in more detail by DAVY in 1812[2]. In spite of the large number of such coordination compounds cited in the literature, it is amazing how few of their properties have been recorded and how little study of their behavior has been pursued.

Conventionally, coordination compounds of an amine with a borane, i. e. amine-boranes, are illustrated by the formula $R{-}N \to B{-}R'$ (with R, R, R on nitrogen and R', R' on boron). These compounds are obtained through direct combination of the components in equimolar amounts with or without the use of solvents at low temperatures.

$$R_3N + BR_3' \to R_3N \cdot BR_3' \qquad (I\text{-}1)$$

However, treatment of an acceptor molecule BR_3 with an electron donor does not always yield a true molecular addition compound in which bonding is effected by the free electron pair of the donor nitrogen. Furthermore, a donor-acceptor linkage of this type possesses considerable polarity and this situation can favor ionization of the molecule or the intramolecular elimination of small molecules from the complex. Consequently, the intramolecular decomposition of amine-boranes leading to aminoboranes, $R_2N{-}BR_2'$, and borazines, $({-}BR'{-}NR{-})_3$ is one of the best known reactions of the amine-borane system. Reactions effected

---

[1] GAY-LUSSAC, J. L., and J. L. THENARD: Mem. de phys. et de chim. de la soc. d'arcueil **2**, 210 (1809).

[2] DAVY, J.: Ann. Chimie **86**, 178 (1813).

through cleavage of the boron-nitrogen linkage, e.g. reactions with suitable Lewis bases, are of equal importance. An example is illustrated in the following equation.

$$\text{LiH} + (\text{CH}_3)_3\text{N} \cdot \text{BH}_3 \rightarrow \text{Li}[\text{BH}_4] + \text{N}(\text{CH}_3)_3 \qquad \text{(I-2)}$$

The following description of amine-boranes will not be confined to the addition products of simple nitrogen bases with a BR_3 acceptor. In view of the varying degree of stability residing in the basic amine-borane system, it seems reasonable to include those addition compounds which are, in general, derived by the coordination of a free electron pair of a nitrogen atom to a BR_3 molecule. Hence, borane complexes with nitriles, amides and related substances will also be considered if sufficient evidence suggests dative boron-nitrogen bonding in the resultant materials.

2. The Normal Coordination Structure (Donor-Acceptor Bonding)

The nature of the boron-nitrogen bond in an amine-borane, formed by donation of the free electron pair of a tervalent nitrogen to a boron component has not been established unequivocally and may vary between two extremes. Incomplete sharing of the electron pair results in a very weak bond whereas complete sharing involves a charge transfer to the acceptor atom resulting in a very strong bond. SIDGWICK[1] suggested that this type of bond should be represented by an arrow as illustrated in I, but the use of charges II, implying an electron-transfer process[2], has received more general acceptance.

The principle of such donor-acceptor bonding has been expressed by MULLIKEN[3, 4] in quantum-mechanical terms. Thus for the highly polar trimethylamine-trifluoroborane, $(\text{CH}_3)_3\text{N} \cdot \text{BF}_3$, b^2 is much greater than a^2 in the wave function, ψ_N, of the ground state.

$$\psi_\text{N} \approx a\,\psi_0[(\text{CH}_3)_3\text{N} \cdot \text{BF}_3] + b\,\psi_1[(\text{CH}_3)_3\text{N} \cdot \text{BF}_3] \qquad \text{(I-3)}$$

In this equation ψ_0 is the no-bond wave function, and ψ_1 is the donor wave function corresponding to a complete transfer of an electron from the nitrogen in trimethylamine to the boron in trifluoroborane in forming a bond by the odd electrons in $(\text{CH}_3)_3\overset{\oplus}{\text{N}} \cdot$ and $\cdot \overset{\ominus}{\text{BF}_3}$. In the more stable amine-boranes of similar type, the second term, which involves ψ_1, predominates. In general, the nature of the donor-acceptor linkage

[1] SIDGWICK, N. V.: The Electronic Theory of Valence. Oxford: Clarendon Press 1927.
[2] LOWRY, T. M.: Trans. Faraday Soc. **18**, 285 (1923).
[3] MULLIKEN, R. S.: J. physic. Chem. **56**, 801 (1952).
[4] MULLIKEN, R. S.: J. Amer. chem. Soc. **74**, 811 (1952).

may vary between the two extremes cited above. The bond strength is affected by the nature of the groups attached to both the boron and the nitrogen atom and consequently, the stability of an amine-borane (which is directly dependent upon the nature of the bond) can vary tremendously. Some of these compounds may be stable only at low temperatures whereas others of comparable size can be distilled without decomposition even at atmospheric pressures.

The stability of selected systems of amine-boranes has been studied frequently in order to establish underlying principles. The classic work of BURG and GREEN[1] provides a relatively well understood example of polar effects. In the series $(CH_3)_3N \cdot BF_3$, $(CH_3)_3N \cdot BF_2CH_3$, $(CH_3)_3N \cdot BF(CH_3)_2$, $(CH_3)_3N \cdot B(CH_3)_3$, trimethylamine-trifluoroborane is the most stable member: The three highly electronegative fluorine atoms increase the LEWIS acidity of the boron acceptor atom compared to the situation where the boron is bonded to electron-releasing alkyl groups.

However, the stability of an addition compound is not always in agreement with expectations based on electronegativity data. For instance, trifluoroborane has long been considered as the strongest boron acceptor molecule known and the acceptor power of trihalogenoboranes was thought to be in the order $BF_3 > BCl_3 > BBr_3$. More recent work, using pyridine and nitrobenzene as reference bases, indicates a reversed sequence; an explanation for the relative acidity of halogenoboranes can be found in terms of π-bonding between the halogen and boron in BX_3 molecules[2].

It is not surprising that, under these circumstances, various types of bonding have been postulated for the amine-borane system. Besides the normal coordination structure, an ionic type and, in the case of aromatic organic amines, a π-complex model has been considered for those amine-boranes containing boron-halogen linkages. The classical formulation of a normal coordination structure is unquestionably the predominant situation in the amine-borane system. If there is no halogen substituent bonded to the boron, the formation of an ionic structure as discussed below seems to be excluded: The free electron pair of the nitrogen adds to the boron atom establishing a lone-pair electron bond and forming a normal molecular addition compound (I). This same concept of bonding holds true for amine-halogenoboranes. It was developed, in principle, by GOUBEAU and coworkers[3] and was substantiated by GERRARD et al.[4]: As described below, there is some evidence that certain amine-boranes such as dimethylamine-trichloroborane, $(CH_3)_2HN \cdot BCl_3$, might have an ionic (i.e. salt-like) structure. However, a chloroform solution of such materials does not show a detectable conductivity[3]. This situation is in agreement with the behavior of other amine-boranes, such as trimethylamine-borane, $(CH_3)_3N \cdot BH_3$, trimethylamine-trimethylborane, $(CH_3)_3N \cdot B(CH_3)_3$, or trimethylamine-trichloroborane, $(CH_3)_3N \cdot BCl_3$. In contrast, trimethylammonium chloride, $[(CH_3)_3NH]Cl$, as a salt, is

[1] BURG, A. B., and A. A. GREEN: J. Amer. chem. Soc. 65, 1838 (1943).

[2] BROWN, H. C., and R. R. HOLMES: ibid. 78, 2173 (1956).

[3] GOUBEAU, J., M. RAHTZ and H. J. BECHER: Z. anorg. allg. Chem. 275, 161 (1954).

[4] GERRARD, W., M. F. LAPPERT and C. A. PEARCE: J. chem. Soc. (London) 1957, 381.

known to be an electrolyte. Moreover, although aminodimethylborane,

$$\begin{array}{c} H \\ \\ H \end{array}\!\!\!\!\!\! N\!-\!B\!\!\!\!\!\!\begin{array}{c} CH_3 \\ \\ CH_3 \end{array}$$, will add one mole of hydrogen chloride to form am-

monia-dimethylchloroborane[1],

$$\begin{array}{c} H \\ H \\ H \end{array}\!\!\!\!\!\! N \to B\!\!\!\!\!\!\begin{array}{c} CH_3 \\ CH_3 \\ Cl \end{array}$$

subsequent action of HCl will cleave the B—N linkage.

$$\begin{array}{c} H \\ \\ H \end{array}\!\!\!\!\!\! N\!-\!B\!\!\!\!\!\!\begin{array}{c} CH_3 \\ \\ CH_3 \end{array} \xrightarrow{HCl} \begin{array}{c} H \\ H \\ H \end{array}\!\!\!\!\!\! N \to B\!\!\!\!\!\!\begin{array}{c} CH_3 \\ CH_3 \\ Cl \end{array} \xrightarrow{HCl} NH_4Cl + Cl\!-\!B\!\!\!\!\!\!\begin{array}{c} CH_3 \\ \\ CH_3 \end{array} \qquad (I\text{-}4)$$

Furthermore, ammonia-dimethylchloroborane cannot add acceptors such as trimethylborane, $B(CH_3)_3$, or trichloroborane, BCl_3, as one would expect in the case of a salt structure[2]. GERRARD and coworkers[3] observed molar conductivities in the order of $\lambda_m \approx 0.2\,\Omega^{-1}\,cm.^{-2}$ for triethylamine-trichloroborane, $(C_2H_5)_3N \cdot BCl_3$, and diethylamine-tri-chloroborane, $(C_2H_5)_2HN \cdot BCl_3$, which is definitely much too low to conform with an ionic structure. Under these circumstances, there does not seem to be a rigorous requirement for the existence of an ionic structure. This view is substantiated by spectroscopic data, primarily since the amine structure of the nitrogen base as indicated by infrared absorptions remains intact when the amine is converted into an amine-borane[4]. Formation of an ionic nitrogen, however, should result in detectable changes evidenced by the observation of the ammonium vibrational increment of the spectrum.

3. The Ionic Complex

The formation of an ionic complex in the amine-borane system was formulated by BROWN and OSTHOFF[5]. Their study of the vapor density of dimethylamine-trichloroborane,

$$\begin{array}{c} H_3C \\ H_3C \\ H \end{array}\!\!\!\!\!\! N \to B\!\!\!\!\!\!\begin{array}{c} Cl \\ Cl \\ Cl \end{array},$$

seems to indicate the presence of a gas phase equilibrium:

$$(CH_3)_2HN \cdot BCl_3 \rightleftharpoons (CH_3)_2N\!-\!BCl_2 \cdot HCl \rightleftharpoons (CH_3)_2N\!-\!BCl_2 + HCl \qquad (I\text{-}5)$$

Therefore, the structure of the amine-borane might equally be written as a salt (III). This view has received some support since the addition

[1] WIBERG, E., and K. HERTWIG: Z. anorg. allg. Chem. **255,** 141 (1947).
[2] GOUBEAU, J., M. RAHTZ and H. J. BECHER: ibid. **275,** 161 (1954).
[3] GERRARD, W., M. F. LAPPERT and C. A. PEARCE: J. chem. Soc. (London) 1957, 381.
[4] KREUTZBERGER, A., and F. C. FERRIS: J. org. Chemistry **27,** 3496 (1962).
[5] BROWN, C. A., and R. C. OSTHOFF: J. Amer. chem. Soc. **74,** 2340 (1952).

of pyridine to a hot solution of dimethylamine-trichloroborane provides

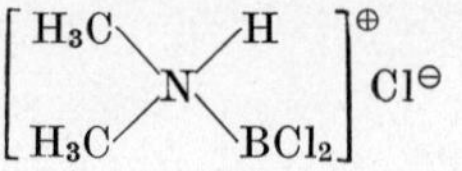

$$\left[\begin{array}{c} H_3C \diagdown \qquad \diagup H \\ N \\ H_3C \diagup \qquad \diagdown BCl_2 \end{array}\right]^{\oplus} Cl^{\ominus}$$

III

an addition complex described as

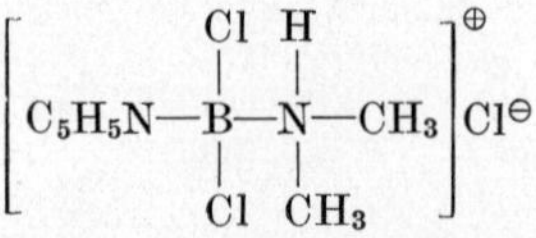

$$\left[\begin{array}{c} \quad Cl \quad H \\ \quad | \quad \ | \\ C_5H_5N-B-N-CH_3 \\ \quad | \quad \ | \\ \quad Cl \quad CH_3 \end{array}\right]^{\oplus} Cl^{\ominus}$$

Furthermore, dimethylamine-trichloroborane is readily hydrolized when placed in an aqueous solution of silver nitrate and yields an immediate precipitate of AgCl. Such behavior does indeed point to an ionic structure of this compound, especially since its rapid hydrolysis is in sharp contrast to the behavior of the rather difficultly hydrolyzed trimethylamine-trichloroborane, $(CH_3)_3N \cdot BCl_3$. It is possible that a rapidly established equilibrium involving the slightly ionized dimethylamine derivative (III) might exist. There are no infrared data available for this particular compound but KREUTZBERGER and FERRIS[1] have demonstrated that, at least in amine-boranes derived from several aromatic diamines, the ammonium vibrational element is not present, thus ruling strongly against an ionic structure in these cases.

4. The π-Complex Model

GERRARD and MOONEY[2] studied a number of amine-boranes which were obtained through the interaction of trifluoroborane or trichloroborane and primary aromatic amines. Largely on the basis of infrared spectral data, these authors postulated that, in such compounds, a π-type bonding might exist. The authors interpret the observation of only one N—H stretching band near 3200 cm.$^{-1}$ as evidence for the structure indicated below (IV). This structure was formulated on the assumption that the classical coordination structure should exhibit at least two N—H bands arising from symmetric and asymmetrical stretching modes of the N—H bond. Furthermore, the similarity of these spectra with those of (arylamino)dihalogenoboranes, $ArHN$—BX_2, in the 850 cm.$^{-1}$ region was thought to lend credence to the indicated structure (IV).

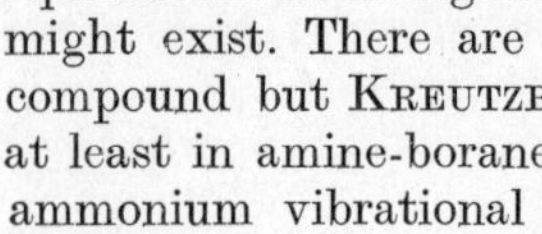

Additional evidence for the π-type structure was provided by ultraviolet spectra, in which λ_{max} did not shift to a longer wavelength

[1] KREUTZBERGER, A., and F. C. FERRIS: J. org. Chemistry 27, 3496 (1962).
[2] GERRARD, W., and E. F. MOONEY: J. chem. Soc. (London) 1960, 4028.

as would be expected for a normal coordination complex in resonance
with a p-quinonoid structure:

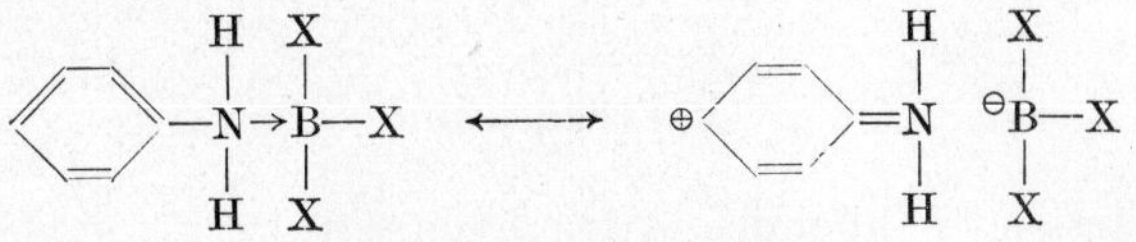

$$
\underset{H \quad X}{\overset{H \quad X}{\bigcirc\!\!\!\!-N\!\rightarrow\!B\!-\!X}} \longleftrightarrow \overset{\oplus}{\bigcirc}=\underset{H \quad X}{\overset{H \quad X}{N \quad \overset{\ominus}{B}\!-\!X}} \tag{I-6}
$$

Some chemical evidence appears to be more questionable: On
reacting aniline-trifluoroborane with phenylmagnesium bromide, (phenylamino)diphenylborane, $(C_6H_5)HN—B(C_6H_5)_2$, alone was obtained. This
behavior was considered to be consistent with the existence of a π-com-
plex structure. However, the course of this reaction does not appear
to be unusual. Intramolecular elimination of HX from an intermediate
might be postulated equally well as the aforementioned suggestion,
and might even be favored if only for steric reasons. Under these circumstances, the following reaction sequence can be formulated:

$$
ArH_2N \cdot BX_3 + ArMgX \rightarrow ArH_2N \cdot BX_2Ar + MgX_2 \tag{I-7}
$$

$$
ArH_2N \cdot BX_2Ar + ArMgX \rightarrow ArH_2N \cdot BXAr_2 + MgX_2 \tag{I-8}
$$

$$
ArH_2N \cdot BXAr_2 \rightarrow ArHN—BAr_2 + HX \tag{I-9}
$$

In addition, KREUTZBERGER and FERRIS[1] reported some spectral
data from complexes of trifluoroborane with aromatic diamines. The
utilization of highly halogenated hydrocarbons as dispersing agents while
the infrared spectra were being taken gave clear evidence for the presence
of two N—H bands as required by a normal coordination structure.
If the boron atom does not have halogen substituents, the formation
of a π-complex structure as formulated by GERRARD seems unlikely.
Consequently, KREUTZBERGER and FERRIS[1] prepared a number of
complexes from triphenylborane and the corresponding amines and
recorded their infrared spectra. The general spectral pattern set by
trifluoroborane with the same amines was found unaltered. Under these
circumstances, it seems unreasonable to consider the similarity of certain
regions of amine-borane spectra with those of aminoboranes as sufficient
evidence for the existence of a π-bonded structure in amine-boranes.
Considering the numerous possibilities for varying the substituents
at both the boron and the nitrogen atom, it is apparent that a single
illustration might not adequately describe the bonding situation in
amine-boranes. Nevertheless, reasonable doubt exists as to the formulation of a π-bond model for the bonding in arylamine-halogenoboranes.
Moreover, NÖTH and BEYER[2] have recently shown that slight impurities can obscure the characteristics of amine-boranes. For instance, it
was long believed that alkylamine-boranes, $RH_2N \cdot BH_3$, are thermally
rather unstable. Actually, however, instability and lowered melting points
of the amine-boranes are caused by traces of impurities.

---

[1] KREUTZBERGER, A., and F. C. FERRIS: J. org. Chemistry **27**, 3496 (1962).
[2] NÖTH, H., and H. BEYER: Chem. Ber. **93**, 928 (1960).

B. Amine-Boranes Derived from $BH_{3-n}R_n$

1. Preparation

The reaction of diborane with ammonia is very complex and can lead to a variety of products depending on the conditions of the reaction. Therefore, this case will be discussed separately (see I-C). When amines react with diborane at low temperatures, symmetrical cleavage of the boron hydride occurs and an amine-borane is readily formed.

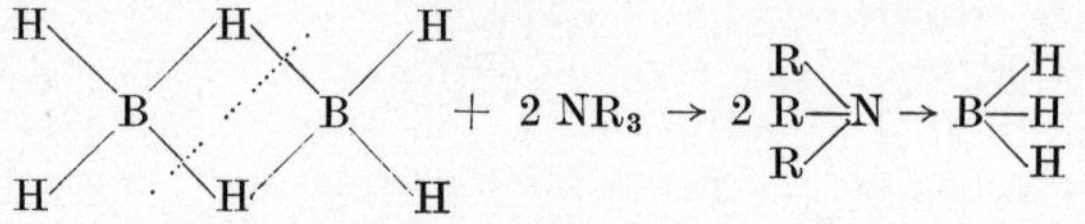

$$\text{(diagram)} \quad + \; 2\,NR_3 \;\rightarrow\; 2\; R{-}N{\rightarrow}B{-}H \qquad (I\text{-}10)$$

This reaction has been studied with a wide variety of cyclic and acyclic primary, secondary and tertiary amines, and higher boron hydrides in place of the diborane. Of necessity, high vacuum techniques have been utilized in this procedure. An important breakthrough was realized when SCHAEFFER and ANDERSON[1] introduced lithium hydroborate as the source of BH_3. In addition, it was also found possible to use amine salts in this reaction:

$$[R_3NH]X + LiBH_4 \rightarrow R_3N \cdot BH_3 + LiX + H_2 \qquad (I\text{-}11)$$

Equally facile syntheses of amine-boranes involving the high pressure hydrogenolysis of trialkylboranes in the presence of amines[2] or the

$$R_3N \cdot BR'_3 + 3\,H_2 \rightarrow R_3N \cdot BH_3 + 3\,R'H \qquad (I\text{-}12)$$

reduction of alkoxyboranes with aluminium and hydrogen in an amine solvent[3] have been effected. Recently it was found[4] that amines react with trihalogenoboranes in the presence of metal hydroborates to yield amine-boranes according to the equation

$$4\,R_3N + BX_3 + 3\,Na[BH_4] \rightarrow 4\,R_3N \cdot BH_3 + 3\,NaX \qquad (I\text{-}13)$$

A representative listing of some alkylamine-borane derivatives of BH_3 is compiled in Table I-1.

As expected, organic diamines will form 1:2 adducts with amine-borane linkages, since both amino groups are available for coordination. Structural problems do not normally appear to exist in these cases. Therefore, it is of interest to consider the diborane adduct of ethylene-diamine. This material was first obtained through the interaction of

---

[1] SCHAEFFER, G. W., and E. R. ANDERSON: J. Amer. chem. Soc. **71**, 2143 (1949).

[2] KÖSTER, R.: Angew. Chem. **69**, 64 (1957).

[3] ASHBY, E. C., and W. E. FOSTER: J. Amer. chem. Soc. **84**, 3407 (1962).

[4] LANG, K., and F. SCHUBERT: U.S. Patent 3,037,985 (1962).

Table I-1. *Alkylamine-Boranes Derived from* BH_3[1]

	m.p., °C	b.p. (mm.), °C
$CH_3NH_2 \cdot BH_3$	56	
$C_2H_5NH_2 \cdot BH_3$	19	
$n\text{-}C_3H_7NH_2 \cdot BH_3$ . . .	45	
$i\text{-}C_3H_7NH_2 \cdot BH_3$ . . .	65	
$n\text{-}C_4H_9NH_2 \cdot BH_3$ . . .	-48	
$t\text{-}C_4H_9NH_2 \cdot BH_3$ . . .	96	
$(CH_3)_2NH \cdot BH_3$ . . .	37	$49\ (10^{-2})$
$(C_2H_5)_2NH \cdot BH_3$ . . .	-18	84 (4)
$(n\text{-}C_3H_7)_2NH \cdot BH_3$ . .	30	
$(i\text{-}C_3H_7)_2NH \cdot BH_3$ . . .	23	88 (1)
$(n\text{-}C_4H_9)_2NH \cdot BH_3$.	15	
$(i\text{-}C_4H_9)_2NH \cdot BH_3$ . . .	19	
$(CH_3)_3N \cdot BH_3$	93.5	
$(C_2H_5)_3N \cdot BH_3$	-2	$42\ (10^{-4})$
$(n\text{-}C_3H_7)_3N \cdot BH_3$ . . .	18	
$(n\text{-}C_4H_9)_3N \cdot BH_3$ . . .	-28	$80\ (10^{-5})$

ethylenediamine with diborane[2] and an open-chain structure has been assigned to the product (V).

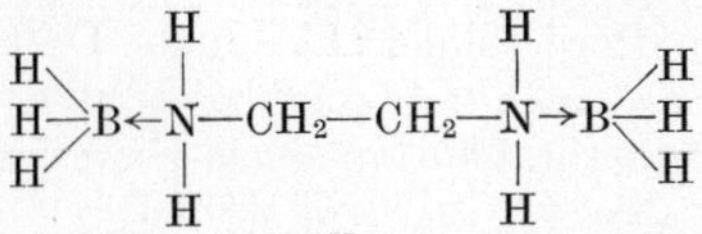

$$\begin{array}{c} H \qquad\qquad H \\ | \qquad\qquad | \\ H{-}B{\leftarrow}N{-}CH_2{-}CH_2{-}N{\rightarrow}B{-}H \\ | \qquad\qquad | \\ H \qquad\qquad H \end{array}$$
V

Largely on the basis of infrared spectral data, a salt-like structure (VI) has been attributed to a material of identical analytical composition which was prepared from sodium hydroborate and ethylenediamine hydrochloride in tetrahydrofuran solution[3].

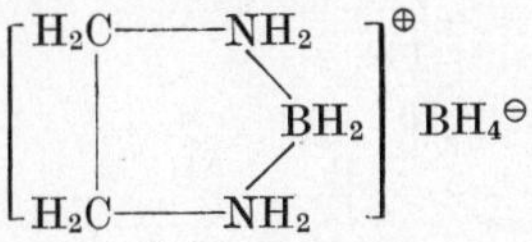

$$\left[\begin{array}{c} H_2C{-}{-}{-}{-}NH_2 \\ \qquad\qquad\qquad BH_2 \\ H_2C{-}{-}{-}{-}NH_2 \end{array}\right]^{\oplus} BH_4^{\ominus}$$
VI

On the basis of molecular weight determinations and ^{11}B nuclear magnetic resonance spectroscopy, KELLY and EDWARDS[4] demonstrated that both materials are identical and have an open-chain structure. Only one boron resonance split into a quadruplet having peaks of relative intensity $1:3:3:1$ could be recorded. This indicates that the material consists of two equivalent boron atoms, each of which is bonded to three hydrogens. In addition, the coupling constant J_{BH} of 88 c.p.s. is in agreement

[1] NÖTH, H., and H. BEYER: Chem. Ber. **93**, 928 (1960).
[2] KELLY, H. C., and J. O. EDWARDS: J. Amer. chem. Soc. **82**, 4842 (1960).
[3] GOUBEAU, J., and H. SCHNEIDER: Chem. Ber. **94**, 816 (1961).
[4] KELLY, H. C., and J. O. EDWARDS: Inorg. Chem. **2**, 226 (1963).

with J_{BH} values reported for other amine-boranes[1] thus providing conclusive evidence for structure V.

The interaction of diborane with hydrazine was originally studied by EMELEUS and STONE[2] but no conclusive data were obtained. Later STEINDLER and SCHLESINGER found that diborane and hydrazine will react to form the adduct $N_2H_4 \cdot 2BH_3$, m.p. 0.4°, at −80° in ether[3]. The hydrazine is readily displaced from the complex by trimethylamine. The adduct $N_2H_4 \cdot BH_3$, m.p. 61°, was obtained through the reaction of hydrazine sulfate with sodium hydroborate[4]; $N_2H_4 \cdot 2BH_3$ also results from this reaction. The pyrolysis of hydrazine-borane was studied in detail; it yielded polymeric products having a backbone of rings consisting of two boron and four nitrogen atoms.

Little information is available on the reaction of diborane with nitriles. BROWN and SUBBA RAO[5] have shown that diborane may function as a reagent for the reduction of nitriles to the corresponding primary amines. The formation of a solid adduct with acetonitrile, $CH_3CN \cdot BH_3$, has also been reported[6]. At room temperature, this adduct reverts to the starting materials accompanied by slight decomposition. A more detailed investigation of this reaction has illustrated that decomposition can also occur with transfer of hydrogen, thereby providing for the formation of borazines[7] (see Chapter III). Sterically encumbered boranes, such as tert. butylborane[8], $C_4H_9BH_2$, or tetraalkyldiboranes[9], $(R_2BH)_2$, react with nitriles to afford solid 1:1 adducts. On the basis of molecular weight determinations, such adducts have been expressed as cyclic dimers (VII). This structural formula has been confirmed in the sense

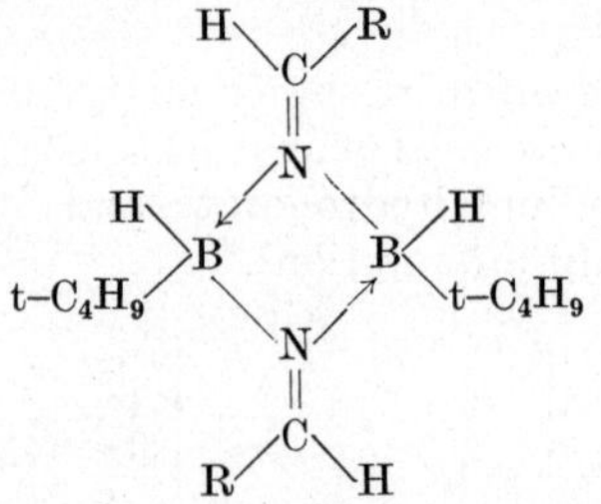

VII

that an analogous compound, Cl_3C—CH=N—BH_2, m.p. 116°, was obtained when trichloroacetonitrile was reacted with diborane in dimethoxy-

---

[1] PHILLIPS, W. D., H. C. MILLER and E. L. MUETTERTIES: J. Amer. chem. Soc. 81, 4496 (1959).

[2] EMELEUS, H. J., and F. G. A. STONE: J. chem. Soc. (London) 1951, 840.

[3] STEINDLER, M. J., and H. I. SCHLESINGER: J. Amer. chem. Soc. 75, 756 (1953).

[4] GOUBEAU, J., and E. RICHTER: Z. anorg. allg. Chem. 310, 123 (1961).

[5] BROWN, H. C., and B. C. SUBBA RAO: J. Amer. chem. Soc. 82, 681 (1960).

[6] SCHLESINGER, H. I., and A. B. BURG: Chem. Revs. 31, 1 (1942).

[7] EMELEUS, H. J., and K. WADE: J. chem. Soc. (London) 1960, 2614.

[8] HAWTHORNE, M. F.: Tetrahedron 17, 117 (1962).

[9] LLOYD, J. E., and K. WADE: J. chem. Soc. (London) 1964, 1649.

ethane[1]. Since $Cl_3C\text{---}CH\!\!=\!\!N\text{---}BH_2$ contains only two equivalents of hydridic hydrogen, the proposed structure seems to be substantiated.

Organoboranes, $BR_{3-n}H_n$, react with either ammonia or amines yielding the respective amine-boranes. A vast number of resultant 1:1 addition products have been reported, as is illustrated by a compilation of some alkylamine-trimethylboranes in Table I-2.

Table I-2. *Alkylamine-Trimethylboranes*

	m.p., ^{0}C	References
$CH_3NH_2 \cdot B(CH_3)_3$ . .	27	2, 3, 4
$(CH_3)_2NH \cdot B(CH_3)_3$ . .	35	2
$(CH_3)_3N \cdot B(CH_3)_3$ . . .	128	2, 5–10
$C_2H_5NH_2 \cdot B(CH_3)_3$ . .	24	3, 11, 12
$(C_2H_5)_2NH \cdot B(CH_3)_3$ . .	27	11, 12
$(C_2H_5)_3N \cdot B(CH_3)_3$ . .	−18 to −14	12
n-$C_3H_7NH_2 \cdot B(CH_3)_3$.	6	3, 13, 14
i-$C_3H_7NH_2 \cdot B(CH_3)_3$.	9	15, 16
n-$C_4H_9NH_2 \cdot B(CH_3)_3$.	3.5	3, 13, 14
s-$C_4H_9NH_2 \cdot B(CH_3)_3$.	−7	15
t-$C_4H_9NH_2 \cdot B(CH_3)_3$.	17 to 19	15
$\begin{array}{c}H_2C\!\!\diagdown\\ \quad\quad\!\!>\!\!NH \cdot B(CH_3)_3 \text{ . .}\\ H_2C\!\!\diagup\end{array}$	10 to 12	17
$\begin{array}{c}H_2C\!\!\diagdown\\ \quad\quad\!\!>\!\!N\text{---}CH_3 \cdot B(CH_3)_3\\ H_2C\!\!\diagup\end{array}$	94.2 to 94.4	18

Amine-organoboranes have also been synthesized by other procedures. HAWTHORNE demonstrated that amine-monoalkylboranes are obtained in good yield through the reaction of B-trialkylboroxines, $(\text{---}BR\text{---}O\text{---})_3$,

[1] LEFFLER, A. J.: Inorg. Chem. **3**, 145 (1964).
[2] BROWN, H. C.: J. Amer. chem. Soc. **67**, 378 (1945).
[3] TAFT, R. W.: ibid. **75**, 4231 (1953).
[4] WIBERG, E., and K. HERTWIG: Z. anorg. allg. Chem. **255**, 141 (1947).
[5] BROWN, C. A., and R. C. OSTHOFF: J. Amer. chem. Soc. **74**, 2340 (1952).
[6] BROWN, H. C.: ibid. **67**, 374 (1945).
[7] BROWN, H. C., H. I. SCHLESINGER and S. Z. CARDON: ibid. **64**, 325 (1942).
[8] BROWN, H. C., M. D. TAYLOR and M. GERSTEIN: ibid. **66**, 431 (1944).
[9] SCHLESINGER, H. I., N. W. FLODIN and A. B. BURG: ibid. **61**, 1078 (1939).
[10] BURG, A. B., and A. A. GREEN: ibid. **65**, 1838 (1943).
[11] BROWN, H. C.: ibid. **67**, 1452 (1945).
[12] BROWN, H. C., and M. D. TAYLOR: ibid. **69**, 1332 (1947).
[13] BROWN, H. C., and M. D. TAYLOR: ibid. **66**, 846 (1944).
[14] BROWN, H. C., M. D. TAYLOR and S. SUJISHI: ibid. **73**, 2464 (1951).
[15] BROWN, H. C., and G. K. BARBARAS: ibid. **75**, 6 (1953).
[16] BROWN, H. C., and H. PEARSALL: ibid. **67**, 1765 (1945).
[17] BROWN, H. C., and M. GERSTEIN: ibid. **72**, 2926 (1950).
[18] McLAUGHLIN, D. E., M. TAMRES, S. SEARLES Jr. and F. BLOCK: J. inorg.
nucl. Chem. **18**, 118 (1961).

with lithium aluminium hydride in the presence of amines[1, 2]. Amine-arylboranes of the types $R_3N \cdot BArH_2$ and $R_3N \cdot BAr_2H$ were prepared through reduction of arylhydroxyboranes with $LiAlH_4$ in the presence of tertiary amines[3].

However, not all amines react with organoboranes to produce amine-boranes. Exceptions which have been reported are readily explained in terms of steric hindrance. For example, examination of molecular models of trimethylborane and 2,2'-dimethylpyridine illustrates that coordination of the free electron pair of the nitrogen to boron would involve considerable strain on both systems and the synthesis of the expected adduct would appear to be virtually impossible[4]. In this connection, it is of interest to note that amine-organoboranes have played an important part in the development of the steric strain theories of H. C. Brown to account for the lack of consistency in the strength sequences of Lewis acids and bases. The interested reader is referred to a review of this effort[5, 6] and also to some work on steric effects in displacement reactions[7].

a. General Procedure for the Preparation of Alkylamine-Boranes from Lithium Hydroborate and Alkylammonium Chloride[8]

A three-necked flask, equipped with a stirrer, dropping funnel and $CaCl_2$ drying tube is charged with a slurry of dried and powdered alkylammonium chloride in anhydrous ether. An ethereal solution of lithium hydroborate is slowly added with vigorous stirring in a nitrogen atmosphere. (The molar ratio of $LiBH_4$: $(RNH_3)Cl$ should be about 0.95:1.0.) When the evolution of hydrogen has ceased, the reaction mixture is stirred for an additional three hours in order to complete the reaction. The product is filtered through a fritted disc in a closed system and the ether evaporated under vacuum. The residue of alkylamine-borane is normally of substantial purity. It can be recrystallized at low temperatures or may be distilled in high vacuum.

b. General Procedure for the Preparation of Trimethylamine-Alkylboranes from $LiAlH_4$ and B-Trialkylboroxines[1]

A solution of 3.8 g. of $LiAlH_4$ in 100 ml. of anhydrous ether is prepared under a slow stream of nitrogen in a three-necked flask equipped with a stirrer, dropping funnel and Dry Ice condenser. The solid hydride is stirred in the refluxing ether until solution is complete; the reaction mixture is then cooled to room temperature and is subjected to the addition of 10 gms. of trimethylamine, the calculated amount of B-trialkylboroxine, dissolved in 25 ml. of ether is then added over an hour while the refluxing reaction mixture is stirred vigorously. Upon completion of the addition, the mixture is refluxed for another hour and then cooled in an ice-bath. Exactly 6.5 ml. of water is added slowly with stirring. The mixture is then filtered and the ether stripped off. The residual trimethylamine-alkylborane is purified by molecular distillation or through recrystallization at low temperatures.

1 Hawthorne, M. F.: J. Amer. chem. Soc. **83**, 831 (1961).
2 Hawthorne, M. F.: ibid. **81**, 5836 (1959).
3 Hawthorne, M. F.: ibid. **80**, 4293 (1958).
4 Brown, H. C., and R. B. Johannesen: ibid. **75**, 16 (1953).
5 Brown, H. C.: Rec. Chem. Progr. **14**, 83 (1953).
6 Lappert, M. F.: Chem. Review **56**, 959 (1956).
7 Brown, H. C., and coworkers: J. Amer. chem. Soc. **77**, 1715 ctd. (1955).
8 Nöth, H., and H. Beyer: Chem. Ber. **93**, 928 (1960).

c. Pyridine-Phenylborane[1]

A quantity, 2.5 g., of $LiAlH_4$ is dissolved in 500 ml. of anhydrous ether under an atmosphere of nitrogen. The solution is cooled to $-70°$ an 10 ml. of pyridine added in one portion. A solution of 17.7 g. of (diethoxy)phenylborane in 70 ml. of ether is added slowly with vigorous stirring over one hour. The reaction mixture is allowed to warm slowly to room temperature with stirring. It is cooled with an ice-bath and a solution of 5 ml. of pyridine in 12 ml. of water is added. After vacuum filtration, the solvent is removed under reduced pressure and the cryst-alline residue of pyridine-phenylborane is recrystallized from ether/pentane at $0°$. Yield: 77%, m.p. 83—85°.

2. Properties and Reactions

Recently it has been shown[2] that alkylamine-boranes are thermally and hydrolytically more stable than was formerly believed, provided they are of sufficient purity. Slight impurities cause not only a consider-able decrease in the melting points of the products but also lower the decomposition temperature by a substantial amount.

The kinetics and mechanism of solvolysis of amine-boranes depends upon a variety of factors such as the kind and amount of substitution on the boron and nitrogen, solvent effects, etc[3]. However, there is very little detailed information available. Hydrolysis of methylamine-boranes has been shown to proceed by displacement of a BH_3 group when a proton attacks the amine nitrogen. Acid hydrolysis of trimethylamine-borane shows the reaction to be first order with respect to amine-borane and the concentration of acid and the rate of reaction increases with the ionic strength of solution. On n-propanolysis of pyridine-borane, $C_5H_5N \cdot BH_3$, cleavage of the boron-nitrogen bond is the rate-determining step and pyridine is displaced by the incoming alcohol. The importance of steric factors in the solvolysis was demonstrated by means of substituted pyridines[4-6]. A third type of reaction produces an amine-boronium cation when a proton attacks a B—H bond directly. This was revealed in a study of the hydrolysis of pyridine-phenylboranes in acetonitrile solution[7]. The results are consistent with the existence of a nonlinear transition state involving electrophilic attack of a water proton on the electrons of a B—H bond.

Dipole moments of a variety of amine-boranes have been measured. The data of Table I-3 indicate an apparent decrease of the dipole moment of an alkylamine-borane with increasing chain length.

[1] HAWTHORNE, M. F.: J. Amer. chem. Soc. **80**, 4293 (1958).
[2] NÖTH, H., and H. BEYER: Chem. Ber. **93**, 928 (1960).
[3] KELLY, H. C., F. R. MARCHELLI and M. B. GIUSTO: Abstr. of Papers, 145th National Meeting of the American Chemical Society, New York, N.Y., 1963, p. 5-N.
[4] RYSCHKEWITSCH, G. E., and E. R. BIRNBAUM: J. physic. Chem. **65**, 1087 (1961).
[5] RYSCHKEWITSCH, G. E.: J. Amer. chem. Soc. **82**, 3290 (1960).
[6] RYSCHKEWITSCH, G. E.: Advances in Chemistry **42**, 53 (1964).
[7] HAWTHORNE, M. F., and E. S. LEWIS: J. Amer. chem. Soc. **80**, 4296 (1958).

Table I-3. *Dipole Moments (in Debye) of some Alkyl-amine-Boranes at 25° in Benzene Solution*[1]

R	$RH_2N \cdot BH_3$	$R_2HN \cdot BH_3$	$R_3N \cdot BH_3$
CH_3	5.19	4.87	4.45
$n\text{-}C_3H_7$	4.72	4.55	
$t\text{-}C_4H_9$	4.64		

An important feature of the vibrational spectra of complexes such as amine-boranes is shown by the relation between the frequency and the nature of the dative bond. This quantity may be expected to measure the nature of the bond strength. Since coupling of the B—N stretch with other skeletal motions is to be expected, it is important to study those environmental factors which contribute to coupling in order to establish the validity of spectral ranges. This has been done for very few amine-boranes; the normal range for the B—N stretching frequency in amine-boranes of the type $R_3N \cdot BX_3$ (R=H, D, CH_3; X=H, D, F, Cl, Br) was found to lie between 700 and 800 cm.[-1] [2]. Normal coordinate calculations substantiated the conclusions resulting from experiment and the assignment was confirmed by isotopic effects. This study established that it is unrealistic to assign characteristic B—N frequencies for the above cited amine-boranes if X=F. This is due to the extensive mixing of B—N and B—F stretching modes.

Proton magnetic resonance spectra of some trialkylamine-boranes $R_3N \cdot BX_3$ (X=H, Cl) have been reported. On the basis of chemical shift data, it was concluded that the π-electron transfer from nitrogen to boron is slightly greater than that shown in borazines[3].

Ammonia-trimethylborane, $H_3N \cdot B(CH_3)_3$, reacts with potassium in liquid ammonia to form potassium aminotrimethylborate, $K[H_2N \cdot B(CH_3)_3]$, which contrasts to the reaction of ammonia-trifluoroborane, $H_3N \cdot BF_3$ [4].

$$H_3N \cdot B(CH_3)_3 + e^{\ominus} \rightarrow [H_2N \cdot B(CH_3)_3]^{\ominus} + \tfrac{1}{2}H_2 \qquad (I\text{-}14)$$

$$H_3N \cdot BF_3 + e^{\ominus} \rightarrow H_2N\text{--}BF_2 + F^{\ominus} + \tfrac{1}{2}H_2 \qquad (I\text{-}15)$$

When heated, potassium aminotrimethylborate readily loses two moles of methane to produce a polyanionic salt. Acid hydrolysis is quantitative in accordance with the equation:

$$[H_2N \cdot B(CH_3)_3]^{\ominus} + 2\,HCl \rightarrow NH_4^{\oplus} + 2\,Cl^{\ominus} + B(CH_3)_3 \qquad (I\text{-}16)$$

With sodium hydride, amine-boranes of the type $R_3N \cdot BH_3$ readily yield sodium hydroborate (eq. I-17)[5].

$$R_3N \cdot BH_3 + NaH \rightarrow NaBH_4 + R_3N \qquad (I\text{-}17)$$

Several of the major reactions of amine-boranes, such as their thermal degradation to aminoboranes and borazines, will be discussed in detail

[1] NÖTH, H., and H. BEYER: Chem. Ber. **93**, 939 (1960).

[2] TAYLOR, R. C.: Advances in Chemistry **42**, 59 (1964).

[3] OHASHI, O., Y. KURITA, T. TOTANI, H. WATANABE, T. NAKAGAWA and M. KUBO: Bull. Chem. Soc. Japan **35**, 1317 (1962).

[4] HOLLIDAY, A. K., and N. R. THOMPSON: J. chem. Soc. (London) **1960**, 2695.

[5] KÖSTER, R.: Angew. Chem. **69**, 94 (1957).

in subsequent chapters. However, amine-boranes have found some importance as intermediates in other organoborane syntheses. A noteworthy example of this reaction is the successful hydroboration of olefines to produce trialkylboranes (eq. I-18)[1-3].

$$R_3N \cdot BH_3 + 3 \underset{}{>}C=C\underset{}{<} \rightarrow R_3N + B(-\overset{|}{\underset{|}{C}}-\overset{|}{\underset{|}{C}}-H)_3 \qquad (I\text{-}18)$$

Other applications of amine-boranes involve their ability to reduce carbonyl functions in certain organic molecules such as aldehydes or ketones[4]. The advantage of amine-boranes over and above metal hydroborates resides in the solubility of the amine derivatives in a very wide variety of solvents, thus providing for homogeneous hydrogenation. The reaction of amine-boranes with thioalcohols gives access to mercaptylboranes. Such compounds were virtually unknown prior to the development of this synthesis[5].

$$RBH_2 \cdot N(CH_3)_3 + 2\,R'SH \rightarrow RB(SR')_2 + 2\,H_2 + N(CH_3)_3 \qquad (I\text{-}19)$$

C. The Chemistry of the Diammoniate of Diborane and Related Structures

1. Ammonia-Borane and the Diammoniate of Diborane

As noted by STOCK and KUSS[6] ammonia and diborane combine in an exothermic reaction to afford a mixture of liquid and solid materials. The nature of the products obtained in this reaction is extremely sensitive to variations in the experimental procedures. However, in vacuum at temperatures below $-80°$ and, in the presence of an excess of ammonia, a solid crystalline material analyzing as BNH_6 is obtained[7] if the excess of ammonia is removed below $-80°$. With an excess of diborane, a compound $B_2H_5NH_2$ is produced[8].

Under these circumstances, one would expect the solid material to be the normal coordination compound $H_3N \cdot BH_3$, but the product was shown to have several properties incompatible with this formula. It is nonvolatile and, in liquid ammonia solution, the material lowers the vapor pressure to a degree corresponding to that expected of the dimeric compound $B_2H_6 \cdot 2NH_3$. This solution conducts electricity and, therefore, the product must have an ionic structure. WIBERG[9] proposed the formula

[1] HAWTHORNE, M. F.: J. org. Chemistry 23, 1788 (1958).
[2] ASHBY, E. C.: J. Amer. chem. Soc. 81, 4791 (1959).
[3] KÖSTER, R.: Angew. Chem. 68, 684 (1957).
[4] NÖTH, H., and H. BEYER: Chem. Ber. 93, 1078 (1960).
[5] HAWTHORNE, M. F.: J. Amer. chem. Soc. 83, 1345 (1961).
[6] STOCK, A., and E. KUSS: Ber. dtsch. chem. Ges. 56, 789 (1923).
[7] STOCK, A., and E. POHLAND: ibid. 59, 2210 (1926).
[8] SCHLESINGER, H. I., D. M. RITTER and A. B. BURG: J. Amer. chem. Soc. 60, 2297 (1938).
[9] WIBERG, E.: Ber. dtsch. chem. Ges. 69, 2816 (1936).

$[NH_4]_2^{\oplus\oplus}$ $[B_2H_4]^{\ominus\ominus}$ but, on the basis of its reaction with one equivalent of sodium in liquid ammonia to give one equivalent of hydrogen,
SCHLESINGER and BURG[1] considered the solid material to be $[NH_4]^{\oplus}$
$[H_3B \cdot NH_2 \cdot BH_3]^{\ominus}$. This formulation has long been accepted since
such a structure is in consonance with the material's chemical reactions,
such as its ready conversion into compounds having B—N—B bonds
and its ability to exchange only N—bonded hydrogen with ND_3 [2].

However, R. W. PARRY and his coworkers have presented conclusive
evidence that the solid material obtained in the low-temperature reaction
between excess ammonia and diborane, the so-called diammoniate of
diborane, is truly a boronate with the constitution:

$$\left[\begin{array}{c} H \\ \diagdown \\ H \end{array} B \begin{array}{c} NH_3 \\ \diagup \\ NH_3 \end{array}\right]^{\oplus} [BH_4]^{\ominus}$$

VIII

The major points in support of this view are as follows:
1. The Raman spectrum of a solution of the material in liquid ammonia
shows all the characteristic frequencies of the boronate anion $[BH_4]^{\ominus}$ [3].
2. Addition of magnesium ions to a solution of the material in liquid
ammonia precipitates the salt $[Mg(NH_3)_6]^{\oplus\oplus}$ $[BH_4]_2^{\ominus\ominus}$ [4] which has been
fully characterized.
3. Reaction of a liquid ammonia solution of the diammoniate of
diborane with alkali metals provides the corresponding metal hydroborates[4], such as $Na[BH_4]$.
4. Addition of ammonium halide to a liquid ammonia solution of
$B_2H_6 \cdot 2NH_3$ precipitates $[H_2B(NH_3)_2]^{\oplus}X^{\ominus}$. The chloride and bromide
have been completely characterized by analysis, molecular weight
determination and X-ray crystallography[4, 5].

Moreover, all the experimental results described above in support
of the earlier formulations are satisfactorily explained by that of PARRY.
This extraordinary structural investigation was accomplished by
rigorously studying the preparation of $B_2H_6 \cdot 2NH_3$ [6]. The basic method
of SCHLESINGER and BURG[1] was improved until uniform materials
were obtained. Both the molecular weight and the dipole moment of
the compound have been determined[7, 8].

It has been noted[9] that $NH_4[BH_4]$ decomposes slowly at room
temperature to the diammoniate of diborane. In ethereal solution, the

[1] SCHLESINGER, H. I., and A. B. BURG: J. Amer. chem. Soc. **60**, 290 (1938).
[2] SHORE, S. G., P. R. GIRARDOT and R. W. PARRY: ibid. **80**, 20 (1958).
[3] TAYLOR, R. C., D. R. SCHULTZ and A. R. EMERY: ibid. **80**, 27 (1958).
[4] SCHULTZ, D. R., and R. W. PARRY: ibid. **80**, 4 (1958).
[5] NORDMAN, C. E., and C. R. PETERS: ibid. **81**, 3551 (1959).
[6] SHORE, S. G., and R. W. PARRY: ibid. **80**, 15 (1958).
[7] PARRY, R. W., G. KODAMA and D. R. SCHULTZ: ibid. **80**, 24 (1958).
[8] WEAVER, J. R., S. G. SHORE and R. W. PARRY: J. chem. Physics **29**, 1 (1958).
[9] PARRY, R. W., D. R. SCHULTZ and P. R. GIRARDOT: J. Amer. chem. Soc.
80, 1 (1958).

latter reacts with ammonium chloride and, in the presence of small amounts of ammonia, not only affords $[H_2B(NH_3)_2]^{\oplus}Cl^{\ominus}$ [1] but also $H_3N \cdot BH_3$ [2].

$$[H_2B(NH_3)_2][BH_4] + NH_4Cl \xrightarrow[\substack{\text{ether} \\ \text{solution}}]{NH_3} [H_2B(NH_3)_2]Cl + H_3N \cdot BH_3 + H_2 \quad (I\text{-}20)$$

Ammonia-borane, $H_3N \cdot BH_3$, was also prepared earlier[3] in 45% yield from the reaction:

$$LiBH_4 + NH_4Cl \rightarrow H_3N \cdot BH_3 + LiCl + H_2 \quad (I\text{-}21)$$

In contrast to the diammoniate of diborane, $H_3N \cdot BH_3$ is ether soluble (although on prolonged standing in solution the diammoniate of diborane slowly precipitates); its molecular weight corresponds to the assigned formula, which is supported by crystal structure work[4, 5]. The B—N bond distance in $H_3N \cdot BH_3$ was found to be 1.56 ± 0.05 Å [2]. This compares favorably with the value of 1.58 ± 0.02 Å in $[H_2B(NH_3)_2]Cl$[6] and of 1.581 Å in $H_3N \cdot B_3H_7$[7]. The B—N stretching frequency of ammonia-borane (in liquid ammonia) was recorded at 785 cm.$^{-1}$ [8], and corresponds to a B—N force constant of about 2.79×10^5 dynes/cm.

In this regard, it is of interest to compare the B—N absorption in some isotopic derivatives of ammonia-borane, $H_3N \cdot BH_3$[8].

Table I-4. *Comparison of ν_{BN} in Isotopic Derivatives of Ammonia-Borane*

Compound	ν_{BN} (cm.$^{-1}$)	ν^*/ν	$\sqrt{\mu/\mu^*}$
$H_3N \cdot BH_3$	785	—	—
$H_3N \cdot BD_3$	737	0.94	0.95
$D_3N \cdot BH_3$	754	0.96	0.97
$D_3N \cdot BD_3$	708	0.90	0.91
$D_3N \cdot B^{10}D_3$	713	0.91	0.93

The spectral range of the B—N stretch is in good agreement with the frequencies available for other coordinated boron bonds with ligands such as CO, PF_3 and dimethyl ether. Consequently, the stretching vibration of a coordinative B—N link seems to be positioned at much lower frequencies than was heretofore assumed.[9-11]

[1] SCHULTZ, D. R., and R. W. PARRY: J. Amer. chem. Soc. **80**, 4 (1958).
[2] SHORE, S. G., and R. W. PARRY: ibid. **80**, 8 (1958).
[3] SHORE, S. G., and R. W. PARRY: ibid. **77**, 6084 (1955).
[4] LIPPERT, E. L., and W. N. LIPSCOMB: ibid. **78**, 503 (1956).
[5] HUGHES, E. W.: ibid. **78**, 502 (1956).
[6] NORDMAN, C. E., and C. R. PETERS: ibid. **81**, 3551 (1959).
[7] NORDMAN, C. E., and C. REIMANN: ibid. **81**, 3538 (1959).
[8] TAYLOR, R. C., and C. L. CLUFF: Nature **182**, 390 (1958).
[9] LUTHER, H., D. MOOTZ and F. RADWITZ: J. prakt. Chem. **277**, 242 (1958).
[10] GOUBEAU, J., and H. J. BECHER: Z. anorg. allg. Chem. **268**, 1 (1952).
[11] RICE, B., R. J. GALIANO and W. J. LEHMANN: J. physic. Chem. **61**, 1222 (1957).

It is remarkable, that the monomer $H_3N \cdot BH_3$ is not available as a primary product of the diborane/ammonia interaction. A sodium derivative, $NaH_2N \cdot BH_3$, was reported much earlier[1].

2. Boronium Salts

During the past few years several boron derivatives with salt structures such as that exhibited in the "diammoniate of diborane" have become available. In 1960 NÖTH and BEYER[2, 3] observed that, upon addition of amines to ethereal solutions of chloroborane, H_2BCl, adducts of the composition $H_2ClB \cdot 2 NR_3$ precipitated. A more detailed study[4] revealed that pyridine-chloroborane, $C_5H_5N \cdot BH_2Cl$, monoalkylaminechloroboranes, $RNH_2 \cdot BH_2Cl$, and dialkylamine-chloroboranes, $R_2NH \cdot BH_2Cl$, react with primary or secondary aliphatic amines to form salt-like compounds of the general formula IX. The authors called

$$\left[\begin{array}{c} H \\ \diagdown \\ \diagup \\ H \end{array} B \begin{array}{c} NH_nR'_{3-n} \\ \\ NH_nR_{3-n} \end{array} \right]^{\oplus} Cl^{\ominus} \quad n = 1,2$$

IX

these compounds "borazylammonium salts". Analogous salts are obtained from the reaction of chloroborane-etherate, $H_2ClB \cdot O(C_2H_5)_2$, with amines (IX, $R = R'$) and through the interaction of alkylammonium halides with an amine-borane as illustrated in the following equation:

$$R_2HN \cdot BH_3 + [R'NH_3]X \xrightarrow{-H_2} \left[H_2B \begin{array}{c} NH_2R' \\ \\ NHR_2 \end{array} \right] X \qquad (I\text{-}22)$$

A variety of such salts were prepared by both of the above methods.[4] They are crystalline materials, soluble in water without decomposition, are stable in acid media and their chloroform solutions have the expected conductivity. Proof of the salt-like structure of the materials is furnished by their infrared spectra which clearly show the absence of δ-NH_2 bands.

The formation of such ionic compounds readily explains an earlier observation[5] that dialkylamine-trichloroboranes are dehydrohalogenated

$$R_2HN \cdot BCl_3 \xrightarrow[-HCl]{(C_2H_5)_3N} R_2N\text{---}BCl_2 \qquad (I\text{-}23)$$

with triethylamine but not with pyridine. In the latter case, borazylammonium salt formation occurs rather than intramolecular elimination of hydrogen chloride. Preference for either reaction might be influenced by stability factors, which in turn are dependent mainly upon steric factors.

[1] SCHLESINGER, H. I., and A. B. BURG: J. Amer. chem. Soc. **60**, 290 (1938).
[2] NÖTH, H., and H. BEYER: Chem. Ber. **93**, 1078 (1960).
[3] NÖTH, H., and H. BEYER: ibid. **93**, 2251 (1960).
[4] NÖTH, H., H. BEYER and H. J. VETTER: ibid. **97**, 110 (1964).
[5] BROWN, J. F.: J. Amer. chem. Soc. **74**, 1220 (1952).

Some other "unusual" amine-boranes, formed by the addition of amines to boranes in a $2:1$ ratio, have likewise been interpreted in terms of boronium salt formation. For example, (alkoxy)dichloroboranes, $ROBCl_2$, have been reported to form addition compounds with pyridine in a $1:2$ molar ratio, e.g. $2\ C_5H_5N \cdot ROBCl_2$ [1]. These adducts, just as those of triiodoborane with two moles of pyridine[2] are considered to be analogous salts. The first such formulation dates back to GOUBEAU and ZAPPEL[3]. These authors obtained the compound X from the reaction of hydrogen chloride on 2-methyl-1,3,2-diazaboracyclopentane.

$$
\begin{bmatrix}
H_2C\!-\!\!-\!\!-\!NH_2 \\
\quad\quad\quad\quad\diagdown \!Cl \\
\quad\quad\quad B \\
\quad\quad\quad\quad\diagdown\! CH_3 \\
H_2C\!-\!\!-\!\!-\!NH_2
\end{bmatrix}^{\oplus} Cl^{\ominus}
$$

X

Another borazylammonium salt involving a cyclic system within the cation was recently described by WIBERG and BUCHLER: Upon reacting trifluoroborane with tetrakis(dimethylamino)ethylene, the salt XI was isolated and identified[4]. The ease of its production may be explained

$$
\begin{bmatrix}
(CH_3)_2N\!-\!C\!-\!\!-\!\!-\!N(CH_3)_2 \\
\quad\quad\quad\quad\quad\quad\quad\quad\diagdown\! F \\
\quad\quad\quad\quad\quad\quad B \\
\quad\quad\quad\quad\quad\quad\quad\quad\diagdown\! F \\
(CH_3)_2N\!-\!C\!-\!\!-\!\!-\!N(CH_3)_2
\end{bmatrix}^{\oplus} F^{\ominus}
$$

XI

by the formation of the sterically favored five-membered ring system. DAVIDSON and FRENCH[5] reported the existence of a cation $[pyridine_2B(C_6H_5)_2]^{\oplus}$; on oxidation of pyridine-phenylborane, $C_5H_5N \cdot BH_2C_6H_5$, with iodine in pyridine, the $[pyridine_2BHC_6H_5]^{\oplus}$ ion is formed[6]. MIKHAILOV and coworkers have obtained similar products[7].

It seems reasonable to conclude that upon reacting some halogenoboranes with amines, the formation of borazylammonium salts predominates over the "normal" dehydrohalogenation as illustrated in eq. (I-23). This situation appears to hold true for chloroborane, BH_2Cl, and diphenylchloroborane, $(C_6H_5)_2BCl$, and, to a lesser degree, for phenyldichloroborane, $C_6H_5BCl_2$, trichloroborane and trifluoroborane. The preference for the formation of borazylammonium salts seems to be governed by the stability of the salt, which in turn is influenced largely by steric factors.

[1] BRINDLEY, P. B., W. GERRARD and M. F. LAPPERT: J. chem. Soc. (London) **1957,** 1540.

[2] MUETTERTIES, E. L.: J. inorg. nucl. Chem. **15,** 182 (1960).

[3] GOUBEAU, J., and A. ZAPPEL: Z. anorg. allg. Chem. **279,** 38 (1955).

[4] WIBERG, N., and J. W. BUCHLER: J. Amer. chem. Soc. **85,** 243 (1963).

[5] DAVIDSON, J. M., and C. M. FRENCH: Chem. and Industry **1959,** 750.

[6] DOUGLAS, J. E.: J. Amer. chem. Soc. **84,** 121 (1962).

[7] MIKHAILOV, B. M., N. S. FEDOTOV, T. A. SCHEGOLEWA and W. D. SCHELUDJAKOV: Bull. Acad. Sci. USSR. **145,** 340 (1962).

The pyrolysis of borazylammonium salts of the type $[H_2B(NR_3)_2]X$ has recently been studied[1]. It provides a convenient route to the preparation of amine-boranes of the type $R_3N \cdot BH_2X$. By this method MUETTERTIES and coworkers[2] have been able to obtain trimethylamine-azidoborane, $(CH_3)_3N \cdot BH_2N_3$.

3. Ammonia and Amine Adducts of Carbon Monoxide-Boranes

BURG and SCHLESINGER[3] noted that carbon monoxide-borane, H_3BCO, reacts with trimethylamine to give trimethylamine-borane, $(CH_3)_3N \cdot BH_3$, suggesting a simple base displacement. On the other hand, ammonia yields a *triammoniate* of carbon monoxide-borane, $H_3BCO \cdot 3NH_3$, which on being warmed, liberates one mole of ammonia to yield the solid "*diammoniate* of carbon monoxide-borane".

A study of several methylamine adducts of carbon monoxide-borane[4] established, by analogy, the structure of the diammoniate as an ionic structure, XII.

$$NH_4^{\oplus} \left[\begin{matrix} H_2N \\ \\ H_3B \end{matrix} \diagdown C=O \right]^{\ominus}$$

XII

The evidence used to establish the structure of XII was based upon the study of $H_3BCO \cdot 2NH_2(CH_3)$ as a model. This evidence is as follows:

1. When reacted with metallic sodium in liquid ammonia, one equivalent of hydrogen is liberated. The resulting product has a salt structure and the reaction may be illustrated by the following equation:

$$[H_3NCH_3]^{\oplus} \left[\begin{matrix} H_2N \\ \\ H_3B \end{matrix} \diagdown C=O \right]^{\ominus} + Na \rightarrow \tfrac{1}{2}H_2 + CH_3NH_2 + Na^{\oplus} \left[\begin{matrix} H_2N \\ \\ H_3B \end{matrix} \diagdown C=O \right]^{\ominus} \quad (I\text{-}24)$$

2. In aqueous solution, the compound reacts with sodium tetraphenylborate with anion exchange. The resultant salt of XII is identical to that obtained through the direct interaction of the ammonium salt of XII with sodium in liquid ammonia.

3. The molecular weight determination in aqueous solution is in agreement with that expected from the above formula, XII.

4. The aqueous solution of XII shows an appreciable degree of conductivity as expected of a salt structure.

5. X-Ray diffraction analysis supports the anionic structure as formulated.

It is reasonable to expect that methylamine will behave like ammonia rather than trimethylamine, since intramolecular migration of a methyl

---

[1] MILLER, N. E., and E. L. MUETTERTIES: J. Amer. chem. Soc. 86, 1033 (1964).

[2] MILLER, N. E., B. L. CHAMBERLAND and E. L. MUETTERTIES: Inorg. Chem. 3, 1064 (1964).

[3] BURG, A. B., and H. I. SCHLESINGER: J. Amer. chem. Soc. 59, 780 (1937).

[4] PARRY, R. W., C. E. NORDMAN, J. C. CARTER and G. TERHAAR: Advances in Chemistry 42, 302 (1964).

group during adduct formation would be difficult. Since B_4H_8CO gives adducts and reactions similar to those of H_3BCO[1], XIII was deduced to be the structure of its diammoniate.

$$NH_4^{\oplus} \left[\begin{array}{c} H_2N \\ \diagdown \\ \diagup \\ H_8B_4 \end{array} C{=}O \right]^{\ominus}$$

XIII

In a formal sense, these anionic complexes are analogous to that of ammonium carbamate except that a BH_3 group replaces the iso-electronic oxygen atom in the carbamate anion. Such an analogy appears reasonable. MULLIKEN[2] has suggested that a BH_3 group and an iso-electronic oxygen atom bear the same electronic relation to each other as diborane and an oxygen molecule. The formal analogy is, of course, too restrictive in an absolute sense, but it has provided the basis for rather extensive structural interpretations[3, 4].

D. Adducts of Trihalogenoboranes with Nitrogen Donor Molecules

1. Trifluoroborane

Since trifluoroborane, BF_3, is one of the most readily available compounds of boron, its addition products with nitrogen derivatives have been studied in great detail. Ammonia was found to react in the gas phase with trifluoroborane to afford a 1:1 adduct[5] and the formula $H_3N \cdot BF_3$ was assigned to the compound[6]. The heat of reaction between BF_3 and NH_3 in the gas phase was found to be 42 kcal.[7], but the existence of a diammoniate and a triammoniate of trifluoroborane, reported in the early literature[8-10] could not be confirmed in this work. However, vapor pressure composition data of the system $NH_3/H_3N \cdot BF_3$ revealed the existence of $BF_3 \cdot 2NH_3$, $BF_3 \cdot 3NH_3$ and $BF_3 \cdot 4NH_3$ at low temperatures[11] and it was proposed that the hydrogen atoms in the 1:1 adduct are sufficiently acid to combine with additional ammonia through hydrogen bonding. At 125—150° ammonia-trifluoroborane decomposes

[1] BURG, A. B., and J. R. SPIELMAN: J. Amer. chem. Soc. **81**, 3479 (1959).
[2] MULLIKEN, R. S.: J. chem. Physics **3**, 635 (1935).
[3] KODAMA, G., and R. W. PARRY: J. inorg. nucl. Chem. **17**, 125 (1961).
[4] NORDMAN, C. E.: Acta crystallogr. [Copenhagen] **13**, 535 (1960).
[5] GAY-LUSSAC, J. L., and J. L. THENARD: Mem. de phys. et de chim. de la Soc. d'arcueil **2**, 210 (1809).
[6] MIXTER, W. G.: Amer. Chem. J. **2**, 153 (1880).
[7] BAUER, S. H., G. R. FINLAY and A. W. LAUBENGAYER: J. Amer. chem. Soc. **67**, 339 (1945).
[8] DAVY, J.: Ann. chim. **86**, 178 (1813).
[9] GAY-LUSSAC, J. L.: Gilbert Ann. der Physik **36**, 6 (1810).
[10] RIDEAL, S.: Ber. dtsch. chem. Ges. **22**, 992 (1889).
[11] BROWN, H. C., and S. JOHNSON: J. Amer. chem. Soc. **76**, 1978 (1954).

irreversibly to boron nitride and ammonium fluoroborate[1]; however, the compound is stable to hydrolysis and may even be recrystallized

$$4\,BF_3 \cdot NH_3 \rightarrow BN + 3\,NH_4[BF_4] \qquad\qquad (I\text{-}25)$$

from water. Cryoscopic measurements of its aqueous solutions indicate that $BF_3 \cdot NH_3$ exists (in such solutions) as a monomer and is not appreciably dissociated[2]. The vibrational spectrum of ammonia-trifluoroborane was recorded by GOUBEAU and MITSCHELEN[3] and they concluded that the force constants K_{BF} in the compound are of the same order of magnitude as in the $BF_4^\ominus$ ion. This conclusion indicates that addition of NH_3 to BF_3 decreases the B—F force constant to the value of a single bond[4]. The foregoing reaction is accompanied by a weakening of the N—H bonds of the ammonia in the addition product, as evidenced by a lowering of the valence vibrations when compared to those of free

[1] LAUBENGAYER, A. W., and G. F. CONDIKE: J. Amer. chem. Soc. 70, 2274 (1948).
[2] CONDIKE, G. F., and A. W. LAUBENGAYER: Abstr. of Papers, 111th National Meeting of the American Chemical Society, Atlantic City, N.J., 1947, p. 7-P.
[3] GOUBEAU, J., and H. MITSCHELEN: Z. physik. Chem. (NF) 14, 61 (1958).
[4] GOUBEAU, J., W. BUES and F. W. KAMPMANN: Z. anorg. allg. Chem. 283, 123 (1956).

Table I-5. *1:1 Adducts of Trifluoroborane with Amines*

Adduct	m.p., °C	References
$NH_3 \cdot BF_3$	163	1, 4, 8, 10, 13, 16
$(CH_3)_3N \cdot BF_3$	145—146	3, 7, 15, 22
$C_2H_5NH_2 \cdot BF_3$	89	11
$(C_2H_5)_2NH \cdot BF_3$	160	11
$(C_2H_5)_3N \cdot BF_3$	29.5	11
$t\text{-}C_4H_9NH_2 \cdot BF_3$	75	17
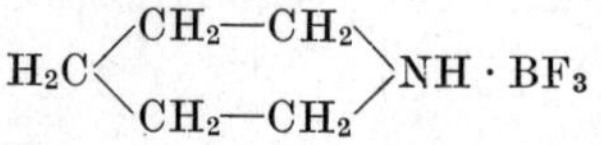 $H_2C{<}^{CH_2-CH_2}_{CH_2-CH_2}{>}NH \cdot BF_3$		2
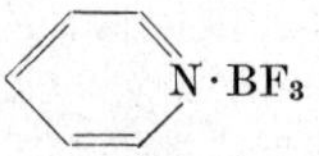 (pyridine) $N \cdot BF_3$	45	22
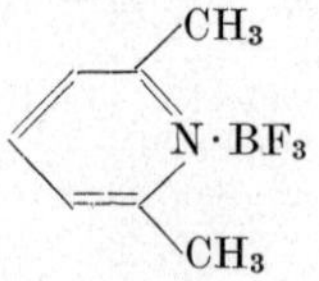 (2,6-dimethylpyridine) $N \cdot BF_3$		6
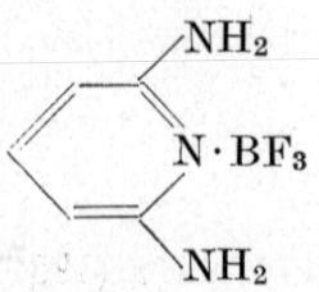 (2,6-diaminopyridine) $N \cdot BF_3$	205—207	12

Adduct	m.p., °C	References
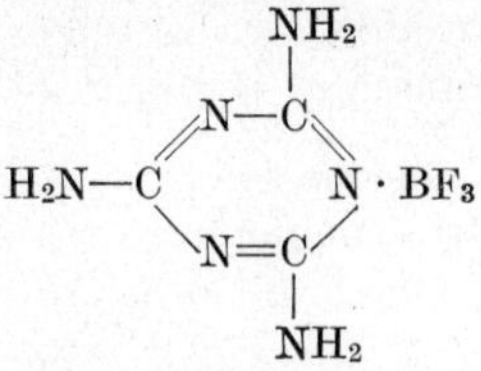 NH_2 structure (H_2N-C ... $N \cdot BF_3$ ... $N=C$ ... NH_2)	420	12
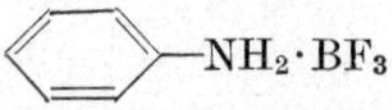 $\bigcirc$–$NH_2 \cdot BF_3$	164—166	9, 14, 19, 21
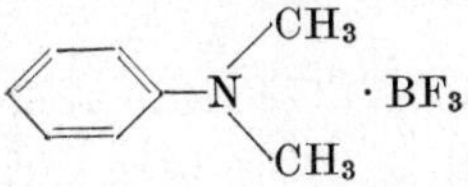 $\bigcirc$–$N(CH_3)_2 \cdot BF_3$	90—92	20, 21
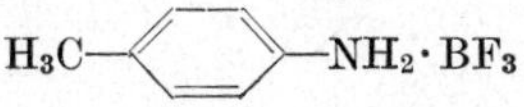 H_3C–$\bigcirc$–$NH_2 \cdot BF_3$	145—150	9, 21
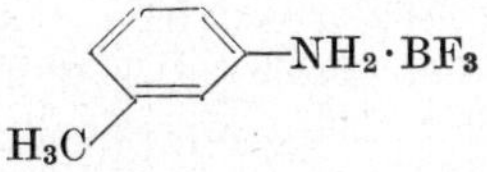 $\bigcirc$–$NH_2 \cdot BF_3$ (H_3C)	130—140	9
$H_2N-CH_2-CH_2-NH_2 \cdot BF_3$	70—74	5
$H_2N-NH_2 \cdot BF_3$	87	18
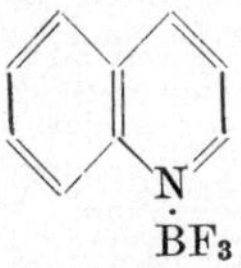 quinoline $\cdot BF_3$		2

[1] BAUER, S. H., G. R. FINLAY and A. W. LAUBENGAYER: J. Amer. chem. Soc. 67, 339 (1945).
[2] BOWLUS, H., and J. A. NIEUWLAND: ibid. 53, 3835 (1931).
[3] BRIGHT, J. R., and W. C. FERNELIUS: ibid. 65, 735 (1943).
[4] BROWN, H. C., and S. JOHNSON: ibid. 76, 1978 (1954).
[5] BROWN, C. A., E. L. MUETTERTIES and E. G. ROCHOW: ibid. 76, 2537 (1954).
[6] BROWN, H. C., H. I. SCHLESINGER and S. Z. CARDON: ibid. 64, 325 (1942).
[7] BURG, A. B., and A. A. GREEN: ibid. 65, 1838 (1943).
[8] GAY-LUSSAC, J. L., and J. L. THENARD: Mem. de phys. et de chim. de la soc. d'arcueil 2, 210 (1809).
[9] GERRARD, W., and E. F. MOONEY: J. chem. Soc. (London) 1960, 4028.
[10] GOUBEAU, J., and H. MITSCHELEN: Z. physik. Chem. (NF) 14, 61 (1958).
[11] KRAUS, C. A., and E. H. BROWN: J. Amer. chem. Soc. 51, 2690 (1929).
[12] KREUTZBERGER, A., and F. C. FERRIS: J. org. Chemistry 27, 3496 (1962).
[13] LAUBENGAYER, A. W., and G. F. CONDIKE: J. Amer. chem. Soc. 70, 2274 (1948).
[14] LANDOLPH, F.: Ber. dtsch. chem. Ges. 12, 1578 (1879).
[15] MEERWEIN, H., E. BATTENBERG, H. GOLD, E. PFEIL and G. WILLFANG: J. prakt. Chem. 154, 83 (1939).
[16] MIXTER, W. G.: Amer. chem. J. 2, 153 (1880).
[17] NÖTH, H., and H. BEYER: Chem. Ber. 93, 2251 (1960).
[18] PATERSON, W. G., and M. ONYSZCHUK: Canad. J. Chem. 39, 986 (1961).
[19] RIDEAL, S.: Ber. dtsch. chem, Ges. 22, 992 (1889).
[20] SNYDER, H. R., H. A. KORNBERG and R. J. ROMIG: J. Amer. chem. Soc. 61, 3556 (1939).
[21] SUGDEN, S., and M. WALOFF: J. chem. Soc. (London) 1932, 1492.
[22] VAN DER MEULEN, P. A., and H. H. HELLER: J. Amer. chem. Soc. 54, 4404 (1932).

ammonia. The lower bond order of N—H in the adduct is also illustrated by the ease with which ammonia-trifluoroborane reacts with alkali metals. This latter reactions was first cited by KRAUS and BROWN[1] and later was studied in more detail[2]. The reaction with potassium is illustrated by the equation:

$$H_3N \cdot BF_3 + K \rightarrow H_2N—BF_2 + KF + \tfrac{1}{2}H_2 \qquad (I-27)$$

When sodium metal is used, a more complex situation arises. The end products might be considered as ammoniates of a hypothetical compound $(H_2N)_2B—NH—B{=}NH$, which could be formed from the above intermediate $H_2N—BF_2$.

Most of the amine substituted addition products of trifluoroborane and organofluoroboranes may be considered as derivatives of the ammonia-trifluoroborane. Some representative adducts of trifluoroborane with amines are listed in Table I-5.

The bond distances of some amine-boranes derived from trifluoroborane have been determined by HOARD and coworkers. It is of interest to note that in $CH_3CN \cdot BF_3$, the B—F bond distance is shortened to 1.32 Å, whereas the B—N bond distance is lengthened to 1.64 Å[3, 4]. The nitrogen of acetonitrile has less donor power than does that of an alkylamine; consequently, the boron-nitrogen linkage is weakened (i.e. lengthened) and accordingly $CH_3CN \cdot BF_3$ is easily hydrolyzed.

Table I-6. *B—N and B—F Bond Distances in Some Amine-Trifluoroboranes*

Compound	B—N distance, Å	B—F distance, Å	References
$H_3N \cdot BF_3$	1.60	1.38	4, 5
$CH_3H_2N \cdot BF_3$	1.57	1.38	4, 6
$(CH_3)_3N \cdot BF_3$	1.59	1.39	4, 7

Although the adduct of pyridine with trifluoroborane is said to be less stable than the corresponding trimethylamine derivative[8], pyridine-trifluoroborane may be distilled without decomposition[9]. The thermal dissociation of this compound was studied and the resultant data are shown in Table I-7.

[1] KRAUS, C. A., and E. H. BROWN: J. Amer. chem. Soc. **51**, 2690 (1929).
[2] McDOWEL, W. J., and C. W. KEENAN: ibid. **78**, 2065 (1956).
[3] HOARD, J. L., T. B. OWEN, A. BUZZELL and O. N. SALMON: Acta Crystallogr. (London) **3**, 130 (1950).
[4] HOARD, J. L., S. GELLER and T. B. OWEN: ibid. **4**, 405 (1951).
[5] HOARD, J. L., S. GELLER and W. N. CASHIN: ibid. **4**, 396 (1951).
[6] GELLER, S., and J. L. HOARD: ibid. **3**, 121 (1950).
[7] GELLER, S., and J. L. HOARD: ibid. **4**, 399 (1951).
[8] BROWN, H. C., H. I. SCHLESINGER and S. Z. CARDON: J. Amer. chem. Soc. **64**, 325 (1942).
[9] VAN DER MEULEN, P. A., and H. H. HELLER: ibid. **54**, 4404 (1932).

From these data, the heat of dissociation for all components in the gaseous state is calculated to be -50.6 kcal.

Aliphatic and aromatic polyamines normally coordinate each amino group with one BF_3 molecule. The first concrete example of the addition

Table I-7. *Dissociation of Pyridine-Trifluoroborane*

Temperature, °C	Apparent Mol. Wt.	%, Degree of Dissociation
356	89.9	63.3
333	108.2	35.7
313	124.1	18.2

of trifluoroborane to polyamines was reported by BURG and MARTIN[1]. They found that hexamethylenetetramine combined with trifluoroborane to form the compound $(CH_2)_6N_4 \cdot 4BF_3$. In ethereal solution, ethylenediamine gave only a 1:1 adduct, $H_2N(CH_2)_2NH_2 \cdot BF_3$, whereas in tetrahydrofuran, the expected 1:2 product, $BF_3 \cdot H_2N(CH_2)_2NH_2 \cdot BF_3$, was obtained[2]. Similarly, hexamethylenediamine reacted with trifluoroborane in tetrahydrofuran to give $BF_3 \cdot NH_2(CH_2)_6NH_2 \cdot BF_3$. This compound is more stable towards hydrolysis than $BF_3 \cdot NH_2(CH_2)_2NH_2 \cdot BF_3$. It was concluded that the weakening effect of one semi-polar B—N bond upon the other (either by a space effect, through the chain, or both) is most significant when the connecting carbon chain is short.

In this connection, it is worth noting the reaction of trifluoroborane with heterocyclic polyamines. Although the polyamines have two or more nitrogen atoms available for coordination, normally only 1:1 adducts are obtained[3]. It would appear that addition occurs at the site of an exocyclic nitrogen and, after the addition of one BF_3 molecule, the unshared electron pair of the other amino nitrogen is not available for reactions of the electrophilic type.

$$\text{(I-28)}$$

The opposite case is illustrated by the behavior of boroxines towards amines[4, 5]. Only one amine molecule is added by a boroxine (XIV), although three boron atoms are incorporated into the ring structure. This appears to indicate that,

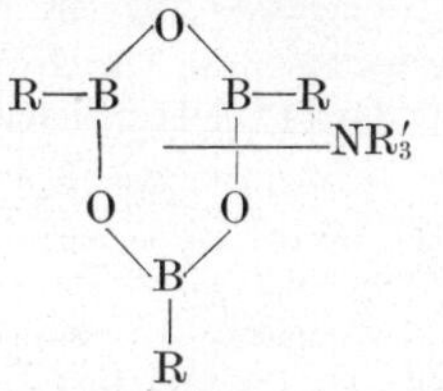

XIV

after one boron is complexed with amine, the remaining two have sufficient electron density to resist the formation of coordination bonds. In the carbocyclic series

[1] BURG, A. B., and L. L. MARTIN: J. Amer. chem. Soc. **65**, 1635 (1943).
[2] BROWN, C. A., E. L. MUETTERTIES and E. G. ROCHOW: ibid. **76**, 2537 (1954).
[3] KREUTZBERGER, A., and F. C. FERRIS: J. org. Chemistry **27**, 3496 (1962).
[4] BURG, A. B.: J. Amer. chem. Soc. **62**, 2228 (1940).
[5] SNYDER, H. R., M. S. KONECKY and W. J. LENNARZ: ibid. **80**, 3611 (1958).

of polyamines, however, every amino group seems to be available for coordination with trifluoroborane.

Table I-8. *Diamine Adducts with Two BF_3 Molecules*

	m.p., °C	References
$BF_3 \cdot H_2N—NH_2 \cdot BF_3$	260	1
$BF_3 \cdot H_2N(CH_2)_2NH_2 \cdot BF_3$	169—170	2
$BF_3 \cdot H_2N(CH_2)_6NH_2 \cdot BF_3$	179—180	2
$BF_3 \cdot H_2N$⟨ ⟩$NH_2 \cdot BF_3$	235—237	3
$BF_3 \cdot H_2N$⟨ ⟩⟨ ⟩$NH_2 \cdot BF_3$	335—336	3
(naphthalene)$—NH_2 \cdot BF_3$ $—NH_2 \cdot BF_3$	272—274	3

A number of 1:1 complexes of nitriles and trifluoroborane have been reported. The acetonitrile derivative has been cited above and shows the boron-nitrogen linkage in these compounds to be considerably weakened as compared to that in the products derived from an aliphatic or aromatic amine. Consequently, the nitrile is readily displaced from these adducts by an amine. However, molecular weight determinations in benzene indicate that the molecule $CH_3CN \cdot BF_3$ exists and is not appreciably dissociated in dilute solutions. The dipole moment of the compound, as measured in benzene at 25°, was found to be 5.8 Debye and the heat of dissociation (into the gaseous reactants) is 26.5 kcal./mole[4]. Characteristics on both, $HCN \cdot BF_3$ [5, 6] and $C_6H_5CN \cdot BF_3$[5] have been reported.

Very little is known about the reaction of trifluoroborane with amides although LANDOLPH[7] reported a 1:2 adduct of brucine with BF_3. Acetamide was reported to absorb one mole of trifluoroborane to form a viscous liquid, $CH_3CONH_2 \cdot BF_3$, which decomposed on distillation[8]. Similar observations were described for the reaction of formamide with trifluoroborane[9]. Dimethylformamide, however, was found to afford a relatively stable material, $HCON(CH_3)_2 \cdot BF_3$. On reacting urea with trifluoroborane, BECHER[10] obtained a 1:1 adduct, $OC(NH_2)_2 \cdot BF_3$, which is readily hydrolyzed since all amide-BF_3 adducts appear to be

[1] PATERSON, W. G., and M. ONYSZCHUK: Canad. J. Chem. **39**, 986 (1961).
[2] BROWN, C. A., E. L. MUETTERTIES and E. G. ROCHOW: J. Amer. chem. Soc. **76**, 2537 (1954).
[3] KREUTZBERGER, A., and F. C. FERRIS: J. org. Chemistry **27**, 3496 (1962).
[4] LAUBENGAYER, A. W., and D. S. SEARS: J. Amer. chem. Soc. **67**, 164 (1945).
[5] POHLAND, E., and W. HARLOS: Z. anorg. allg. Chem. **207**, 242 (1932).
[6] PATEIN, G.: C. R. hebd. Séances Acad. Sci. **113**, 85 (1891).
[7] LANDOLPH, F.: Ber. dtsch. chem. Ges. **12**, 1578 (1879).
[8] BOWLUS, H., and J. A. NIEUWLAND: J. Amer. chem. Soc. **53**, 3835 (1931).
[9] MUETTERTIES, E. L., and E. G. ROCHOW: ibid. **75**, 490 (1953).
[10] BECHER, H. J.: Chem. Ber. **89**, 1691 (1956).

extremely sensitive to hydrolysis. Thermal decomposition of urea-trifluoroborane above 125° yields ammonium fluoroborate, $[NH_4][BF_4]$, and $(HNCO)_x$. It is not yet established with certainty whether or not the boron is actually bonded to the nitrogen in these complexes.

Analogously, amino alcohols such as ethanolamine, $H_2N—CH_2—CH_2—OH$, form coordination compounds with trifluoroborane[1]. In these cases, the donor could also be the oxygen atom as well as the nitrogen. Some recent nuclear magnetic resonance studies seem to favor B—O bonding, while certain chemical reactions appear to indicate the existence of a B—N linkage. The acetamide complex of trifluoroborane has been shown to provide the corresponding acetate when heated with alcohols[2], whereas with amides, a nitrile is obtained[3].

$$CH_3CONH_2 \cdot BF_3 + ROH \rightarrow CH_3COOR + H_3N \cdot BF_3 \qquad (I\text{-}29)$$

$$CH_3CONH_2 \cdot BF_3 + CH_3CONH_2 \rightarrow CH_3CN + CH_3COOH + H_3N \cdot BF_3 \quad (I\text{-}30)$$

Some other addition compounds of trifluoroborane of the amine-borane type should be described. PATERSON and ONYSZCHUK[4] investigated the reaction of trifluoroborane with hydrazine. In ether as solvent, products were obtained in other than stoichiometric proportion, whereas reaction of the same components in the absence of solvents at 25° gave a 1:1 adduct. In tetrahydrofuran solution, the expected 2:1 adduct was obtained. On hydrolyzing this adduct, the $[BF_4]^{\ominus}$ moiety was observed. Further, thermal decomposition produced a variety of materials such as nitrogen, ammonia, ammonium fluoroborate and boron nitride. At −80°, the 1:1 adduct of hydrazine with trifluoroborane forms a diammoniate which is unstable at higher temperatures; it is probably formed by dipole-dipole interaction and it was not possible to isolate compounds of stoichiometric composition. In aqueous solutions $N_2H_4 \cdot BF_3$ undergoes a disproportionation reaction in the sequence:

$$N_2H_4 \cdot BF_3 + H_2O \rightleftharpoons [N_2H_5]^{\oplus}[BF_3OH]^{\ominus} \qquad (I\text{-}31)$$

$$4[BF_3OH]^{\ominus} \rightarrow 3BF_4^{\ominus} + [B(OH)_4]^{\ominus} \qquad (I\text{-}32)$$

2. Interaction of Nitrogen Oxides and Boranes and Related Chemistry

Trifluoroborane has been reported to give a 1:1 adduct with hydroxylamine[5] but no characterizing data are available. With NOF, trifluoroborane combines to yield $NOBF_4$ which is also obtained through the reaction of N_2O_3 or NO_2 with concentrated aqueous solutions of HBF_4. Finally, a compound $NOF \cdot 2BF_3$ has been obtained from trifluoroborane and NOCl.

[1] MILLER, M. A.: U.S. Patent 2,238,068 (1941).
[2] SOWA, F. J., and J. A. NIEUWLAND: J. Amer. chem. Soc. 55, 5052 (1933).
[3] BROWN, H. C., and R. R. HOLMES: ibid. 78, 2173 (1956).
[4] PATERSON, W. G., and M. ONYSZCHUK: Canad. J. Chem. 39, 986 (1961).
[5] GOUBEAU, J.: FIAT Review of German Sciences, Vol. 23, p. 218.

Apart from the fact that NO forms addition compounds with trihalogenoboranes, reactions of nitrogen oxides with boranes have received little attention. However, recent studies have produced some interesting data. BROIS[1, 2] found that the reaction of NO with trialkylboranes is extremely temperature dependent. He reported that NO adds to BR_3 at $-30°$ in a nucleophilic reaction. The resultant 1:1 addition compound with $N \rightarrow B$ coordination rearranges with the migration of an alkyl group and, in an excess of NO, XV, is obtained.

$$BR_3 + 2NO \xrightarrow{-30°} R_2B-N\begin{smallmatrix}NO\\OR\end{smallmatrix} \qquad (I\text{-}33)$$
$$XV$$

At higher temperatures, the nature of the product indicates the primary coordination of oxygen is to boron and thus the reaction may be expressed as:

$$NO + BR_3 \rightarrow R_2B \leftarrow \underset{R}{O}N \xrightarrow{70°} [R_2B-ONR] \xrightarrow{BR_3} R_2B-ON\begin{smallmatrix}R\\BR_2\end{smallmatrix} + R_2B-ONR_2$$
$$XVI \qquad\qquad XVII$$
$$(I\text{-}34)$$

The competition between the two mechanisms illustrated in the above equations (I-33 and I-34) in this temperature dependent reaction is supported by nuclear magnetic resonance studies. These studies dispute a previously postulated mechanism for a reaction of NO with metalorganics[3] which inferred $O \rightarrow B$ coordination and the subsequent 1.3-shift of an alkyl group to yield $R_2B-O-N\begin{smallmatrix}R\\NO\end{smallmatrix}$. However, the work of BROIS finds additional support through an independent study on the interaction of tri-n-butylborane with NO[4]. At room temperature, products of type XVI and XVII were obtained along with another major product having the formula $(R_2BNO)_2$, XVIII. On the basis of its hydrolytic stability and a remarkable resistance to oxidation by air, the compound was assigned the structure of a B-nitroso dimer;

$$R_2B-NO$$
$$\|$$
$$ON-BR_2$$
$$XVIII$$

Although not conclusive, spectroscopic evidence suggests the presence of a nitroso group as is indicated by medium to strong absorption in the infrared region at 1215 and 1175 cm.$^{-1}$. In the ultraviolet, XVIII shows bands at 293 and 228 mμ, closely resembling the bands assigned to aliphatic C-nitroso dimers.

[1] BROIS, S. J.: Abstr. of Papers 144th National Meeting of the American Chemical Society. Los Angeles, Calif., 1963, p. 32-M.
[2] BROIS, S. J.: Tetrahedron Letters 7, 345 (1964).
[3] ABRAHAM, M. A., J. H. N. GARLAND, J. H. HILL and L. F. LARKWORTHY: Chem. and Ind. 1962, 1615.
[4] INATOME, M., and L. P. KUHN: Advances in Chemistry 42, 183 (1964).

It is reasonable to consider the aforementioned products of the interaction of NO with trialkylboranes as derivatives of hydroxylamine. This view is supported by the fact that a similar derivative, XIX, is obtained from the interaction of hydroxylamine with dialkylhydroxyboranes.

$$R_2BOH + HON\!\!< \quad \to \quad R_2B\!-\!O\!-\!N\!\!< + H_2O \qquad (I\text{-}35)$$
$$XIX$$

The reaction of a N-substituted hydroxylamine was shown to yield a monomer of type XIX[1]; with hydroxylamine, dimeric products are obtained which probably have the indicated cyclic structure, XX.

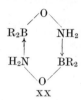

XX

It is worth noting that the oxidation of decahydrodecaborate salts[2, 3] with mild oxidants such as ferric nitrate was accompanied by the appearance of an intense blue color suggesting the presence of a nitroso or nitro group. Indeed, it was found that this reaction essentially parallels that of the decahydrodecaborate ion; for example, the reaction of bistriethylammonium decahydrodecaborate, $B_{10}H_{10}[HN(C_2H_5)_3]_2$, with nitrogen dioxide, results in the formation of a nitrosohydroborate, $B_{20}H_{18}NO[HN(C_2H_5)_3]_3$ [4]. The nitrosation occurs before coupling to the B_{20} unit, since $B_{20}H_{18}^{\ominus\ominus}$ does not form a nitroso derivative with NO_2. The composition of the nitrosohydroborate was established by X-ray diffraction[5]. The infrared spectrum of the compound exibited a strong absorption near 1180 cm.$^{-1}$, which was assigned to the NO mode. The nitroso group is readily reduced by a Raney nickel—catalyzed hydrogenation to yield $B_{20}H_{18}NH_2^{\ominus\ominus\ominus}$ and nearly all the hydrogens of nitrosoborane can be substituted by halogens to a varying degree[4].

3. Trichloroborane Derivatives

Early reports on the formation of addition compounds of ammonia and trichloroborane in various stoichiometric ratios[6-8], $xNH_3 \cdot yBCl_3$,

[1] KUHN, L. P., and M. INATOME: J. Amer. chem Soc. 85, 1206 (1963).
[2] HAWTHORNE, M. F., and A. R. PITOCHELLI: ibid. 81, 5519 (1959).
[3] LIPSCOMB, W. N., A. R. PITOCHELLI and M. F. HAWTHORNE: ibid. 81, 5833 (1959).
[4] WIESBOECK, R. A.: ibid. 85, 2725 (1963).
[5] KACZMARCZKY, A., R. D. DOBROTT and W. N. LIPSCOMB: Proc. Nat. Acad. Sci. USA 48, 729 (1962).
[6] BERZELIUS, J. J.: Ann. Physik 78, 113 (1824).
[7] MARTIUS, C. A.: Liebigs Ann. Chem. 109, 79 (1859).
[8] BESSON, A.: C. R. hebd. Séances Acad. Sci. 110, 516 (1890).

seem to be rather questionable. This discrepancy was first noted by JOANNIS who studied the interaction of the two materials under various experimental conditions[1]. It is very likely that ammonolysis of the boron-halogen linkage occurs rather than coordination. The primary reaction thought to occur is illustrated by the equation:

$$BCl_3 + 6NH_3 \rightarrow B(NH_2)_3 + 3NH_4Cl \qquad (I\text{-}36)$$

The trisaminoborane, $B(NH_2)_3$, is unstable and has never been characterized. It decomposes with the elimination of ammonia and the formation of a polymeric material having the composition $[B_2(NH)_3]_x$.

Substituted ammonias, however, have been found to yield isolable products with trichloroborane. Thus a variety of BCl_3 addition compounds with amines have been reported, some of which are listed in Table I-9.

Table I-9. *BCl_3 Addition Compounds with Amines*

Adduct	m.p., °C	References
$(CH_3)_3N \cdot BCl_3$	243	6
$(C_2H_5)_3N \cdot BCl_3$	92—93.5	2
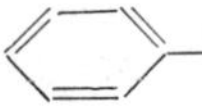 ⬡—$NH_2 \cdot BCl_3$	140	4
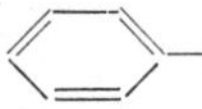 ⬡—$N(CH_3)_2 \cdot BCl_3$	146	3, 4
H_3C—⬡—$NH_2 \cdot BCl_3$ 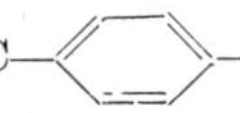	159—160	3, 5
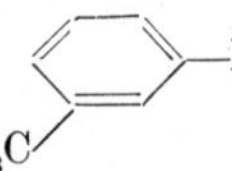 ⬡—$NH_2 \cdot BCl_3$ (with H_3C)	130—140	3
Br—⬡—$NH_2 \cdot BCl_3$ 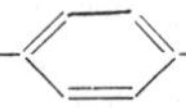	300	3
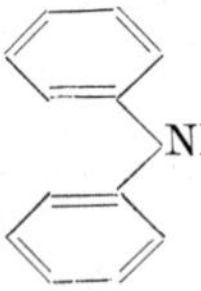 ⬡⬡$NH \cdot BCl_3$		3

All of these adducts are susceptible to hydrolysis and have a distinct tendency to decompose at elevated temperatures yielding various borazines (see Chapter III).

[1] JOANNIS, A.: C. R. hebd. Séances Acad. Sci. **135**, 1106 (1902).
[2] GERRARD, W., M. F. LAPPERT and C. A. PEARCE: J. chem. Soc. (London) **1957**, 381.
[3] GERRARD, W., and E. F. MOONEY: ibid. **1960**, 4028.
[4] JONES, R. G., and C. R. KINNEY: J. Amer. Chem. Soc. **61**, 1378 (1939).
[5] KINNEY, C. R., and M. J. KOLBEZEN: ibid. **64**, 1584 (1942).
[6] WIBERG, E., and W. SÜTTERLIN: Z. anorg. allg. Chem. **202**, 31 (1931).

The compound $H_3N \cdot BH_2Cl$ has been reported to be an unstable material[1]. However, the corresponding trimethylamine adduct was isolated on treatment of $(CH_3)_3N \cdot BH_3$ with anhydrous HCl gas[2].

$$(CH_3)_3N \cdot BH_3 + HCl \rightarrow (CH_3)_3N \cdot BH_2Cl + H_2 \qquad (I\text{-}37)$$

This reaction is of general utility for the preparation of amine-mono-chloroboranes of the type $R_3N \cdot BH_2Cl$ even at low temperatures[3]. HBr and HI react in an analogous manner to that of HCl. In contrast to this behavior, ammonia-trimethylborane reacts with anhydrous HCl through cleavage of the coordinate $B \leftarrow N$ link (eq. I-38)

$$H_3N \cdot B(CH_3)_3 + HCl \rightarrow NH_4Cl + B(CH_3)_3 \qquad (I\text{-}38)$$

affording trimethylborane and ammonium chloride[4]. At $100°$ $(CH_3)_3N \cdot BH_2Cl$ reacts with an excess of hydrogen chloride through additional exchange of hydrogen and formation of $(CH_3)_3N \cdot BHCl_2$. Other exchange reactions of B-bonded hydrogen with halogen have been found possible through the use of trihalogenoboranes[3]; trifluoroborane appears to be the exception, since with trifluoroborane or hydrogen fluoride, amine-boranes of the type $BH_3 \cdot NR_3$ exchange all three hydrogens simultaneously, even at very low temperatures[5]. It is quite possible that a partially fluorinated derivative is first formed. However, this compound is probably extremely unstable and immediately dispro-portionates. Apparently this tendency for disproportionation is not so evident in the case of the higher halogenoboranes.

$$3 R_3N \cdot BH_2F \rightarrow 2 R_3N \cdot BH_3 + R_3N \cdot BF_3 \qquad (I\text{-}39)$$

Trialkylamine-monochloroboranes, $R_3N \cdot BH_2Cl$, have also been obtained on direct halogenation of amine-boranes of the type $R_3N \cdot BH_3$. Another method of preparation involves the reaction of the etherate of BH_2Cl[6] with amines in a simple displacement reaction.

$$R_2O \cdot BH_2Cl + NR_3' \rightarrow R_3'N \cdot BH_2Cl + R_2O \qquad (I\text{-}40)$$

However, with pyridine as a tertiary base, an adduct $BH_2Cl \cdot 2$ pyr. is obtained. A variety of other tertiary amines, ammonia and some primary and secondary amines also react to produce analogous $1:2$ adducts as discussed above.

Amine-monohalogenoboranes are crystalline materials. In the case of primary or secondary amines, the compounds are thermally quite unstable (formation of borazines!) and are sensitive towards hydrolysis. Table I-10 lists some of the amine-monohalogenboranes, $R_3N \cdot BH_2X$, which have recently become available by one of the above described methods.

[1] STOCK, A., and E. POHLAND: Ber. dtsch. chem. Ges. **59**, 2210 (1926).

[2] BURG, A. B., and H. I. SCHLESINGER: J. Amer. chem. Soc. **59**, 780 (1937).

[3] NÖTH, H., and H. BEYER: Chem. Ber. **93**, 2251 (1960).

[4] SCHLESINGER, H. I., N. W. FLODIN and A. B. BURG: J. Amer. chem. Soc. **61**, 1078 (1939).

[5] BROWN, H. C., and P. A. TIERNEY: J. inorg. nucl. Chem. **9**, 51 (1959).

Table I-10. *Amine-Monohalogenoboranes*[1]

Adduct	m.p., °C
$CH_3NH_2 \cdot BH_2Cl$	47
$t\text{-}C_4H_9NH_2 \cdot BH_2Cl$	102
$(CH_3)_2NH \cdot BH_2Cl$	18
$(CH_3)_3N \cdot BH_2Cl$	85
$(C_2H_5)_3N \cdot BH_2Cl$	43
$C_5H_5N \cdot BH_2Cl$	45
$CH_3NH_2 \cdot BH_2Br$	10
$t\text{-}C_4H_9NH_2 \cdot BH_2Br$	98
$(CH_3)_2NH \cdot BH_2Br$	5—6
$(CH_3)_3N \cdot BH_2Br$	67
$CH_3NH_2 \cdot BH_2I$	8—9
$t\text{-}C_4H_9NH_2 \cdot BH_2I$	99
$(CH_3)_2NH_2 \cdot BH_2I$	25
$(CH_3)_3N \cdot BH_2I$	73

Reaction of amine-boranes with a deficiency of trichloroborane (or tribromoborane) to give compounds of the type $R_3N \cdot BH_nX_{3-n}$ [1,2] suggests that this reaction proceeds via hydrogen-halogen exchange without rupture of the boron-nitrogen bond. Similarly, interaction of triethylamine-trifluoroborane, $(C_2H_5)_3N \cdot BF_3$, with trichloroborane under mild conditions liberates trifluoroborane without significant B—N bond rupture[3]. Under more vigorous reaction conditions, processes involving exchange of boron atoms bonded to the donor atoms become more important as has been demonstrated with labeled boron derivatives. However, in the reaction of trifluoroborane with trimethylamine complexes of organodichloroboranes to afford organodifluoroboranes, cleavage of the boron-nitrogen bond always occurs accompanied by halogen transfer[4]. In general, amine-organohalogenoboranes, $R_3N \cdot BR'X_2$ and $R_3N \cdot BR'_2X$, are relatively rare. Only a few compounds of such types have been reported. The first to be characterized were trimethylamine-difluoromethylborane, $(CH_3)_3N \cdot BF_2(CH_3)$, and trimethylamine-dimethylfluoroborane, $(CH_3)_3N \cdot BF(CH_3)_2$ [5]. Trimethylamine-dibromoborane, $(CH_3)_3N \cdot BHBr_2$, m.p. 126—127°, has only recently been reported[1].

Several amine adducts of alkylchloroboranes are known. They are, in general, formed either through direct combination of the reactants in an inert solvent or by a reaction analogous to that described above, i.e. from the reaction of hydrogen chloride on amine-alkylboranes.

$$R_3N \cdot BH_2CH_3 + HCl \rightarrow R_3N \cdot BHClCH_3 + H_2 \qquad (I\text{-}41)$$

In addition, the action of hydrogen chloride on aminoboranes has been described as is illustrated in eq. I-42.

$$R_2N\text{—}BR_2 + HCl \rightarrow R_2HN \cdot BR_2Cl \qquad (I\text{-}42)$$

Derivatives of the type $ROBCl_2 \cdot NR_3$ and $(RO)_2BCl \cdot NR_3$ are known, but apparently offer few interesting features[6-8].

[1] Nöth, H., and H. Beyer: Chem. Ber. 93, 2251 (1960).
[2] Ratajzak, S.: Bull. Soc. chim. France 1960, 487.
[3] Coyle, T.D.: Proc. chem. Soc. (London) 1963, 172.
[4] Brinckman, F. E., and F. G. A. Stone: J. Amer. chem. Soc. 82, 6235 (1960).
[5] Burg, A. B., and A. A. Green: ibid. 65, 1838 (1943).
[6] Meerwein, H., and W. Pannwitz: J. prakt. Chem. 141, 123 (1934).
[7] Wiberg, E., and W. Sütterlin: Z. anorg. allg. Chem. 222, 92 (1935).
[8] Abel, E. W., J. D. Edwards, W. Gerrard and M. F. Lappert: J. chem. Soc. (London) 1957, 501.

The addition of hydrogen cyanide to trichloroborane has been studied briefly[1]. The cited reference indicates that a molecular compound of 1:1 ratio, $HCN \cdot BCl_3$, is formed. However, neither potassium nor silver cyanide will react with trichloroborane[2, 3]. The coordinating power of the CN group is increased greatly when it is coupled with organic moieties. Thus acetonitrile forms a 1:1 complex with trichloroborane, $CH_3CN \cdot BCl_3$[4-6]; the heat of dissociation into gaseous components amounts to 33.4 kcal./mole[7].

A number of nitrile adducts, of trichloroborane were recently studied in greater detail[8, 9]. The results indicate that the nitrile group provides electrons to boron more easily than an ether or a nitro group. For example, the reaction of p-methoxybenzonitrile or p-nitrobenzonitrile with trichloroborane results in the formation of a nitrile complex. On the other hand, since nitriles are weaker bases than amines, it is no surprise that the nitrile in such addition compounds is readily replaced by amines such as pyridine (eq. I-43). This behavior is in agreement with the conclusions

$$RCN \cdot BCl_3 + C_5H_5N \rightarrow C_5H_5N \cdot BCl_3 + RCN \qquad (I-43)$$

derived from a study of bond length data derived from the adducts of nitriles with trifluoroborane as reported above[10]. The complexes of organic nitriles with trichloroborane are thermally rather stable. However, at high temperatures either dissociation (elimination of BCl_3) or decomposition (elimination of HCl) occurs and sometimes both. On the basis of infrared spectral data, it was concluded[8, 11] that a normal coordination structure exists in compounds of this type. The $C \equiv N$ stretching frequency of the nitrile is shifted slightly to higher wavelengths by the coordination. If addition across the triple bond had occurred, resulting in structure XXI, one would expect a significant variation in the $C \equiv N$ absorption. By

$$R—CCl{=}N—BCl_2$$
XXI

way of comparison, examination of infrared spectra of trichloroborane adducts with amides clearly reveals that, in such compounds, coordination occurs from oxygen to boron rather than from nitrogen to boron[9], XXII.

[1] MARTIUS, C. A.: Liebigs Ann. Chem. **109**, 79 (1859).
[2] GUSTAVSON, G.: Ber. dtsch. chem. Ges. **3**, 426 (1870).
[3] GUSTAVSON, G.: Z. Chem. **14**, 417 (1871).
[4] NESPITAL, W.: Z. physik. Chem. B **16**, 153 (1932).
[5] ULICH, H., and W. NESPITAL: Angew. Chem. **44**, 750 (1931).
[6] ULICH, H., and W. NESPITAL: Z. Elektrochem. **37**, 559 (1931).
[7] LAUBENGAYER, A. W., and D. S. SEARS: J. Amer. chem. Soc. **67**, 164 (1945).
[8] COERVER, H. J., and C. CURRAN: ibid. **80**, 3522 (1958).
[9] GERRARD, W., M. F. LAPPERT and J. W. WALLIS: J. chem. Soc. (London) **1960**, 2178.
[10] HOARD, J. L., T. B. OWEN, A. BUZZELL and O. N. SALMON: Acta crystallogr. (London) **3**, 130 (1950).
[11] GERRARD, W., M. F. LAPPERT, H. PYSZORA and J. W. WALLIS: J. chem. Soc. (London) **1960**, 2144, 2182.

Pyridine-trichloroborane[1].—Equimolar amounts of pyridine and trichloroborane are combined in pentane solution at $-10°$. The solid precipitate is dissolved in benzene and precipitates as a crystalline powder, m.p. $115°$, upon the addition of pentane.

4. Tribromoborane and Triiodoborane

Amine coordination compounds of tribromoborane have been studied less intensively than those of the fluorine and chlorine derivatives. Ammonia adducts of tribromoborane in 1:1 molar amounts have been reported in the early literature[2] and other entities such as $2\,BBr_3 \cdot 9\,NH_3$[3] and $BBr_3 \cdot 4\,NH_3$[3,4] have been claimed but not confirmed; their existence is just as questionable as those of similar trichloroborane adducts. Later investigations under various reaction conditions have not been able to unequivocally establish the course of the reaction between ammonia and tribromoborane[5,6]. It appears, however, that the formation of addition compounds is highly unlikely to occur in the BBr_3/NH_3 system; rather, it seems more reasonable to assume that, in analogy to the case of trichloroborane, the boron-bromine linkage undergoes ammonolysis[7]. This view is supported by the fact that, even in dilute solutions and under carefully controlled temperatures, no addition compounds of tribromoborane have been found when the borane is added to either methylamine or ethylamine[8]. Rather, partial aminolysis of boron-halogen bonds occurs leading to derivatives such

as (ethylamino)dibromoborane, $\mathrm{H_5C_2{\diagdown}\atop H{\diagup}}N{-}B{{\diagup}Br\atop{\diagdown}Br}$, and bis(methylamino)-

bromoborane, $(CH_3HN)_2BBr$.

$$BBr_3 + 2\,RNH_2 \rightarrow {R{\diagdown}\atop H{\diagup}}N{-}B{{\diagup}Br\atop{\diagdown}Br} + [RNH_3]Br \qquad (I\text{-}44)$$

Details as to the cited formation of dimethylamine-tribromoborane $(CH_3)_2HN \cdot BBr_3$, and diethylamine-tribromoborane, $(C_2H_5)_2HN \cdot BBr_3$, are not available[9]. In general, lower dialkylamines react in a manner identical to that of monoalkylamines, i.e. through aminolysis of the boron-halogen bond. Upon reacting tribromoborane with higher alkylamines, such as iso-amylamine, or aromatic amines, it was found possible to obtain amine-borane structures of the type $RH_2N \cdot BBr_3$. Also, piperidine and a variety of tertiary amines form molecular compounds with tribromo-borane, most of which are quite unstable and decompose with the

[1] GERRARD, W., and M. F. LAPPERT: J. chem. Soc. (London) **1955**, 1020.
[2] POGGIALE, M.: C. R. hebd. Séances Acad. Sci. **22**, 124 (1846).
[3] STOCK, A.: Ber. dtsch. chem. Ges. **34**, 955 (1901).
[4] BESSON, A.: C. R. hebd. Séances Acad. Sci. **114**, 542 (1892).
[5] POHLAND, E.: Z. anorg. allg. Chem. **201**, 282 (1931).
[6] STOCK, A., and W. HOLLE: Ber. dtsch. chem. Ges. **41**, 2095 (1908).
[7] JOANNIS, A.: C. R. hebd. Séances Acad. Sci. **139**, 364 (1904).
[8] JOHNSON, A. R.: J. physic. Chem. **16**, 1 (1912).
[9] GOUBEAU, J.: FIAT Review of German Sciences, p. 219.

elimination of RX and the formation of aminoboranes[1]. In general, data characterizing their structure or identity are not available for the amine-tribromoboranes. The adduct of tribromoborane with trimethylamine, XXIV, m.p. 238°, stands out as the lone exception[2].

$$
\begin{array}{c}
H_3C \\
H_3C - N \rightarrow B - Br \\
H_3C \\
\end{array}
\quad
\begin{array}{c}
Br \\
Br \\
Br
\end{array}
$$

XXIV

When tribromoborane is reacted in the gaseous state with hydrogen cyanide under reduced pressure $HCN \cdot BBr_3$ is obtained; an analogous adduct of acetonitrile has also been described[3]. It is reasonable to assume the formation of a B—N donor-acceptor bond in these adducts. With the exception of trichloroborane, addition compounds of nitriles with trihalogenoboranes are not very plentiful. No derivatives with triiodoborane have yet been reported. The known adducts of trifluoroborane and tribromoborane are combined in Table I-11.

Table I-11. *1:1 RCN Adducts of BF_3 and BBr_3*

$HCN \cdot BF_3$	$HCN \cdot BBr_3$
$CH_3CN \cdot BF_3$	$CH_3CN \cdot BBr_3$
$C_6H_5CN \cdot BF_3$	$C_2H_5CN \cdot BBr_3$
$H_3C-\!\!\!\left\langle\!\!\!\bigcirc\!\!\!\right\rangle\!\!\!-CN \cdot BF_3$	$C_6H_5CN \cdot BBr_3$
	$C_6H_5CH_2CN \cdot BBr_3$
	$AgCN \cdot BBr_3$

It is remarkable that silver cyanide forms a molecular addition compound with tribromoborane but not with trichloroborane or trifluoroborane. Since the product has been characterized as $AgCN \cdot BBr_3$, this might be considered as the first evidence for the greater acceptor power of tribromoborane over that of the other two cited halogenoboranes, a fact which has only recently been accepted.

The knowledge of triiodoborane/amine interactions is rather limited. Amine-boranes derived from BI_3 are not known, although the reaction of triiodoborane with ammonia has been studied. The precipitation of a material $BI_3 \cdot 5NH_3$ upon introducing ammonia into a solution of BI_3 in carbon tetrachloride at 0° has been claimed[4], but most likely the material consists of a mixture of ammonolysis and condensation products. This view was advanced by JOANNIS[5] and was recently confirmed by a more detailed study of the reaction of triiodoborane with liquid ammonia[6].

[1] JOHNSON, A. R.: J. physic. Chem. **16**, 1 (1912).
[2] NÖTH, H., and H. BEYER: Chem. Ber. **93**, 2251 (1960).
[3] POHLAND, E.: Z. anorg. allg. Chem. **201**, 282 (1931).
[4] BESSON, A.: C. R. hebd. Séances Acad. Sci. **114**, 542 (1892).
[5] JOANNIS, A.: ibid. **135**, 1106 (1902).
[6] McDOWELL, W. J., and C. W. KEENAN: J. Amer. chem. Soc. **78**, 2069 (1956).

E. Amine Adducts of -Oxyboranes, Boroxazolidines, and Ring Systems with Annular N→B Bonds

1. Amine Adducts of -Oxyboranes

Hydroxyorganoboranes, R_2BOH and $RB(OH)_2$, have been found to give 1:1 adducts with amines[1-4] and with other nitrogen bases such as hydrazine and hydroxylamine[4]. It is reasonable to assume N→B coordination in such compounds. Moreover, if the organic substituent at the boron atom contains trivalent nitrogen, the latter's free electron pair can also coordinate with the boron to form a cyclic structure. This was recently demonstrated when 2-(2-boronophenyl)-benzimidazole, XXV, and similar compounds were prepared[5].

NH

N

B

HO   OH

XXV

Evidence for the intramolecular boron-nitrogen coordination was provided by infrared spectral data. The absence of strong absorption in the 7.2—7.8 μ region of the dihydroxyborane spectrum is indicative of tetracoordinated boron[6], whereas absorption between 8.2 and 8.8 μ is evidence of amine coordination in such compounds [7, 8].

Aryloxyboranes, the aryl esters of boric acid, readily form adducts with strongly basic amines in similar fashion[9-12]. However, in the simple aliphatic series only (trimethoxy)borane, $B(OCH_3)_3$, is known to form similar complexes[13-18]. [11]B chemical shift data have been related to

[1] YABROFF, D. L., and G. E. K. BRANCH: J. Amer. chem. Soc. **55**, 1663 (1933).
[2] SNYDER, H. R., and C. WEAVER: ibid. **70**, 232 (1948).
[3] GOUBEAU, J., and J. W. EWERS: Z. anorg. allg. Chem. **304**, 230 (1960).
[4] SERAFIN, B., and M. MAKOSZA: Tetrahedron **19**, 821 (1963).
[5] LETSINGER, R. L., and D. B. MacLEAN: J. Amer. Chem. Soc. **85**, 2230 (1963).
[6] BELLAMY, L. J., W. GERRARD, M. F. LAPPERT and R. L. WILLIAMS: J. chem. Soc. (London) **1958**, 2412.
[7] LETSINGER, R. L., and S. B. HAMILTON: J. Amer. chem. Soc. **80**, 5411 (1958).
[8] GREENWOOD, N. N., and K. WADE: J. chem. Soc. (London) **1960**, 113.
[9] COLCLOUGH, T., W. GERRARD and M. F. LAPPERT: ibid. **1955**, 907.
[10] COLCLOUGH, T., W. GERRARD and M. F. LAPPERT: ibid. **1956**, 3006.
[11] ABEL, E. W., W. GERRARD, M. F. LAPPERT and R. SHAFFERMAN: ibid. **1958**, 2895.
[12] MEHROTRA, R. C., and G. SRIVASTAVA: J. Indian chem. Soc. **39**, 526 (1962).
[13] BROWN, H. C., and E. A. FLETCHER: J. Amer. chem. Soc. **73**, 2808 (1951).
[14] URS, S. U., and E. S. GOULD: ibid. **74**, 2948 (1952).
[15] HORN, H., and E. S. GOULD: ibid. **78**, 5772 (1956).
[16] GOUBEAU, J., and U. BOHN: Z. anorg. allg. Chem. **266**, 116 (1951).
[17] GOUBEAU, J., and R. LINK: ibid. **267**, 27 (1952).
[18] GOUBEAU, J., and E. EKHOFF: ibid. **268**, 145 (1952).

the structure of the crystalline adducts[1]. Also, there is some evidence for the formation of unstable adducts of triethoxyborane[1, 2]. Steric effects might be responsible for the failure of higher aliphatic esters of boron to give stable products with amines[3], although back-coordination of electrons from oxygen (eq. I-45) has also been discussed[4]. Cyclic

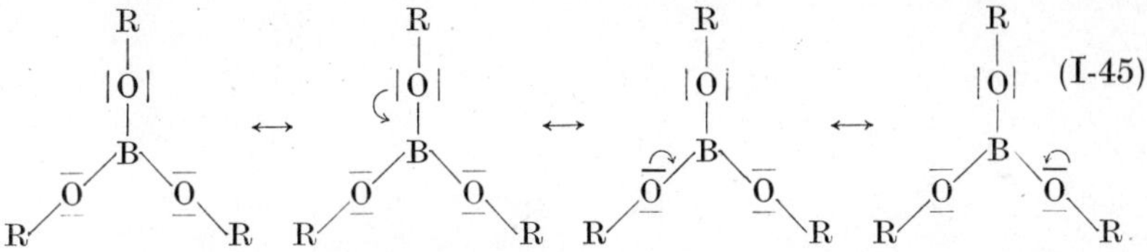

$$
\begin{array}{c}
R \\
| \\
|O| \\
| \\
B \\
O \quad O \\
R \quad R
\end{array}
\longleftrightarrow
\begin{array}{c}
R \\
| \\
|O| \\
| \\
B \\
O \quad O \\
R \quad R
\end{array}
\longleftrightarrow
\begin{array}{c}
R \\
| \\
|O| \\
| \\
B \\
O \quad O \\
R \quad R
\end{array}
\longleftrightarrow
\begin{array}{c}
R \\
| \\
|O| \\
| \\
B \quad O \\
O \quad O \\
R \quad R
\end{array}
\quad \text{(I-45)}
$$

esters of the type XXVI normally give 1:1 adducts with amines; their stability depends upon the size of the ring and the nature of the substituents on the annular carbon atoms[5]. For the amine component, steric considerations are very important, in accordance with H. C. BROWN's theory of base-strength order.

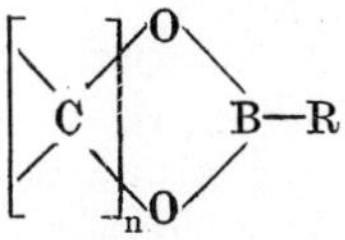

$$
\left[\begin{array}{c} C \end{array}\right]_n
\begin{array}{c} O \\ B-R \\ O \end{array}
$$

XXVI

The esters formed from diorganohydroxyboranes and aminoalcohols, appear to be of special interest. They permit the convenient handling of compounds of the type R_2BOH. LETSINGER and coworkers[6-9] prepared diarylhydroxyboranes as esters by the following method.

$$
B(OR)_3 \xrightarrow{ArMgX} Ar_2BOR \xrightarrow[\text{alcohol}]{\text{amino-}} Ar_2B-O-CH_2CH_2-NH_2 \quad \text{(I-32)}
$$

It is recognized that coordination of the nitrogen's free electron pair to the boron atom, XXVII, contributes to the unusual stability of such compounds.

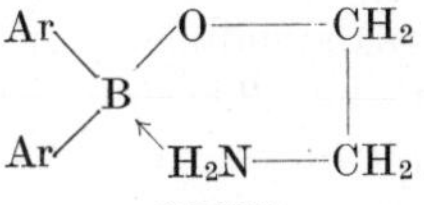

$$
\begin{array}{c}
Ar \quad O-\!\!-\!\!-CH_2 \\
B \quad | \\
Ar \quad H_2N-\!\!-\!\!-CH_2
\end{array}
$$

XXVII

[1] LANDESMAN, H., and R. E. WILLIAMS: J. Amer. chem. Soc. **83**, 2663 (1961).
[2] URS, S. U., and E. S. GOULD: ibid. **74**, 2948 (1952).
[3] BROWN, H. C., and E. A. FLETCHER: ibid. **73**, 2208 (1951).
[4] ABEL, E. W., W. GERRARD, M. F. LAPPERT and R. SHAFFERMAN: J. chem. Soc. (London) **1958**, 2895.
[5] FINCH, A., and J. C. LOCKHART: ibid. **1962**, 3723.
[6] LETSINGER, R. L., I. SKOOG and N. REMES: J. Amer. chem. Soc. **76**, 4047 (1954).
[7] LETSINGER, R. L., and I. SKOOG: ibid. **77**, 2491 (1955).
[8] LETSINGER, R. L., and N. REMES: ibid. **77**, 2489 (1955).
[9] LETSINGER, R. L., and J. R. NAZY: J. org. Chemistry **23**, 914 (1958).

Derivatives with one alkyl and one aryl group attached to the boron have also been prepared[1] and it was found that aromatic alcohols such as 8-quinolinol[1] (as in XXVIII) or pyridine-2-carbinol[2] give even better preparative results.

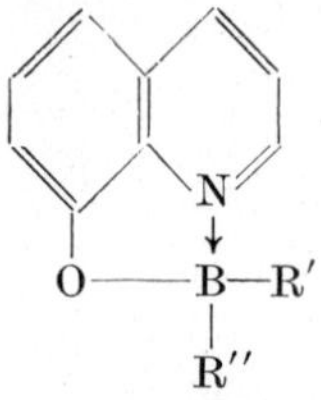

XXVIII

In a similar manner, hydroxylaminoalcohols have been utilized in this reaction and some derivatives of type XXIX have been described[3].

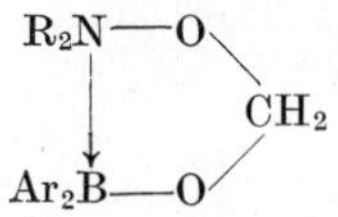

XXIX

2. Boroxazolidines

The boron esters of diethanolamine and triethanolamine are known as boroxazolidines[4].

The first report on a boroxazolidine system is recorded in a German patent in 1933[5], but it was not until 1951 that triethanolamine esters of boron were rediscovered[6]; their preparation was subsequently improved by HEIN and BURCKHARDT[7]. The pioneering preparative work of LETSINGER and his coworkers was extended by H. K. ZIMMERMAN[8-10]; a variety of aminoalcohol derivatives of boron have since become available (see ref. [8] for detailed literature).

Two basic types of boroxazolidines will now be discussed:

(a) Materials derived from 1.3.6.2-dioxazabora-cyclooctane, XXX, in which the free electron pair of the nitrogen atom can coordinate

[1] DOUGLASS, J. E.: J. org. Chemistry 26, 1312 (1961).

[2] NEU, R.: Arch. Pharmaz. Ber. dtsch. pharmaz. Ges. 294, 173 (1961).

[3] ZIMMER, G., and W. RITTER: Angew. Chem. 74, 217 (1962).

[4] The boroxazolidine nomenclature was proposed by H. K. ZIMMERMANN[9, 10] and was based on an earlier suggestion by H. C. BROWN[6]; it is utilized here for convenience; the systematic designations are illustrated below.

[5] ROJAHN, C. A.: German Patent 582,149 (1933).

[6] BROWN, H. C., and E. A. FLETCHER: J. Amer. chem. Soc. 73, 2808 (1951).

[7] HEIN, F., and R. BURCKHARDT: Z. anorg. allg. Chem. 268, 159 (1952).

[8] ZIMMERMAN, H. K.: Advances in Chemistry 42, 23 (1964).

[9] WEIDMANN, H., and H. K. ZIMMERMAN: Liebigs Ann. Chem. 619, 28 (1958).

[10] WEIDMANN, H., and H. K. ZIMMERMAN: ibid. 620, 4 (1959).

to the boron atom thus forming a bicyclic system (diptych boroxazolidine, XXXI).

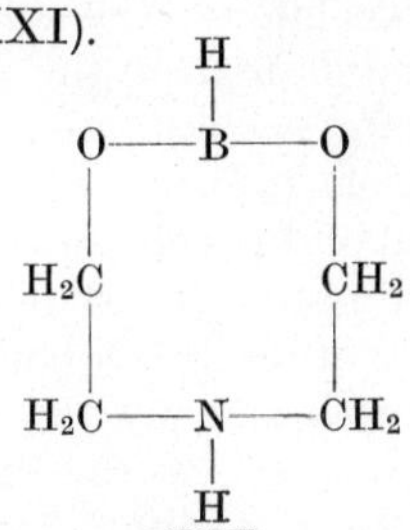

O——B——O
H$_2$C   CH$_2$
H$_2$C——N——CH$_2$
XXX

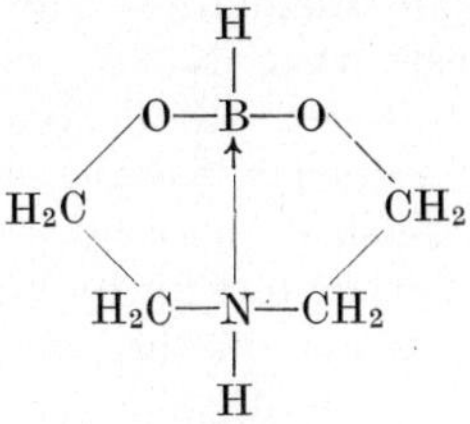

H
O—B—O
H$_2$C   CH$_2$
H$_2$C—N—CH$_2$
H
XXXI

(b) Boron derivatives of triethanolamine (the systematic name of which is 2,8,9,5,1-trioxazabora-bicyclo-[3,3,3]-undecane), XXXII, are known as triptych boroxazolidines. This system is also stabilized by a transannular N→B coordinating bond (XXXIII) analogous to XXXI above.

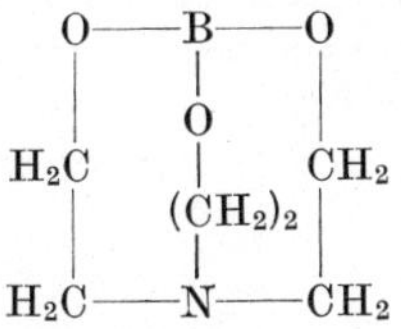

O——B——O
O
H$_2$C   (CH$_2$)$_2$ CH$_2$
H$_2$C——N——CH$_2$
XXXII

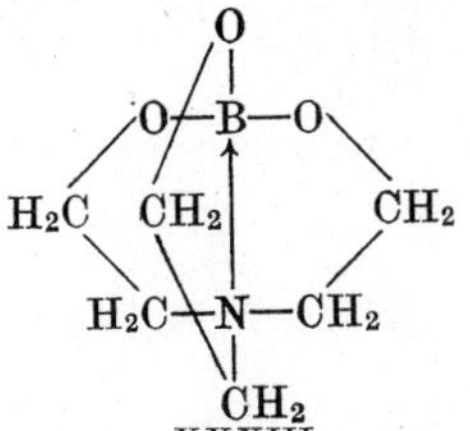

O
O—B—O
H$_2$C   CH$_2$ CH$_2$
H$_2$C—N—CH$_2$
CH$_2$
XXXIII

Support for a N→B coordination structure in such derivatives has been obtained from various data. To begin with, boroxazolidines are very stable towards hydrolysis in contrast to the hydrolytic instability of most normal alkoxyboranes and aryloxyboranes[1]. This fact suggests that the electron deficiency at the boron is satisfied by donation of an electron pair from the nitrogen. In addition, the N—H stretching frequency in diptych boroxazolidines shows a large shift in the 3100 cm.$^{-1}$ region of the infrared spectrum if compared in the same region with that of normal secondary amines. This shift is related to the formation of a tetrahedral nitrogen in the boroxazolidine system, i.e. N→B coordination, although LAWESSON indicated[2] that it is of much higher order (in the order of about 150—200 cm.$^{-1}$) than in the normal amine-boranes. Dipole measurements were not of any real use in discriminating between the two possible structures[3]. However, the results of detailed kinetic studies of triptych boroxazolidines have been found to be consistent with a hydrolysis mechanism involving primary cleavage of a boron-nitrogen coordination bond, followed by the normally rapid ester hydrolysis characteristic of boric esters[4, 5]. The kinetic data are not in

[1] MUSGRAVE, O. C., and T. O. PARK: Chem. and Ind. 1955, 1552.
[2] LAWESSON, S. O.: Ark. Kemi 10, 171 (1956).
[3] FU, H. C., T. PSARRAS, H. WEIDMANN and H. K. ZIMMERMAN: Liebigs Ann. Chem. 641, 116 (1961).
[4] WEIDMANN, H., and H. K. ZIMMERMAN: ibid. 628, 37 (1959).
[5] ZIMMERMAN, H. K.: Advances in Chemistry 42, 23 (1964).

agreement with earlier reasoning by STEINBERG and HUNTER[1] which related the hydrolytic stability solely to steric effects. Indeed, sterically hindered boroxazolidines show an enhanced hydrolytic instability. This suggests that N→B coordination can be inhibited by steric factors[2]. However, the delayed base reaction towards aqueous acids, which was studied in a detailed investigation of diptych boroxazolidines, clearly established the existence of a coordinated N→B linkage. The activation parameters for hydrolytic cleavage of the B—N bond revealed a strong influence on the rate of reaction with acid by the size of the nitrogen-substituent. This can be explained only by the existence of a coordinated boron-nitrogen bond. Additional support for such coordination can be found in the hydrolytic instability of a neutral solution of tris(ethanolamine-borane) borate, $B(OCH_2CH_2NH_2 \cdot BH_3)_3$ [3]. Since the nitrogen in the latter compound is coordinated to BH_3 groups, the central boron atom remains electron deficient and, therefore, promotes the hydrolytic instability of the ester. Furthermore, 6-methyl-2-p-bromophenyl-1,3,6,2-dioxazaboracyclooctane, XXXIV, does not react easily with GRIGNARD reagents[4], whereas normal alkoxyboranes readily form B—C bonds in this reaction[5].

$$
\begin{array}{c}
Br \\
| \\
\text{(C}_6\text{H}_4\text{)} \\
\end{array}
$$

$$
\begin{array}{ccc}
& O\diagdown\ \underset{\uparrow}{B}\ \diagup O & \\
H_2C & & CH_2 \\
CH_2 & N & CH_2 \\
& CH_3 & \\
\end{array}
$$

XXXIV

3. Other Ring Systems with Annular Coordinated N→B Bonding

In addition to the boroxazolidines and similar compounds described above, several other heterocyclic systems have been reported in which ring closure is effected by a coordinated N→B bond. For example, a five-membered ring, XXXV, was obtained through the reaction of dimethylallylamine with trimethylamine-borane in toluene[6].

$$
\begin{array}{c}
CH{=}CH_2 \\
| \\
CH_2 \\
| \\
N(CH_3)_2
\end{array}
\ +\ (CH_3)_3N\cdot BH_3 \longrightarrow
\begin{array}{c}
CH_2 \\
\diagup \diagdown \\
CH_2\quad CH_2 \\
|\qquad | \\
(CH_3)_2N\longrightarrow BH_2
\end{array}
\ +\ N(CH_3)_3 \qquad (I\text{-}46)
$$

XXXV

[1] STEINBERG, H., and D. L. HUNTER: Ind. Engng. Chem. **49**, 174 (1957).
[2] SCHLEPPNIK, A. A., and C. D. GUTSCHE: J. org. Chemistry **27**, 3684 (1962).
[3] KELLY, H. C., and J. O. EDWARDS: J. Amer. chem. Soc. **82**, 4842 (1960).
[4] HOFFMANN, A. K., and W. M. THOMAS: ibid. **81**, 580 (1959).
[5] BEAN, F. R., and J. R. JOHNSON: ibid. **54**, 4415 (1932).
[6] ADAMS, R. M., and F. D. POHOLSKY: Inorg. Chem. **2**, 640 (1963).

A cyclic system, XXXVI, in which the central boron atom exhibits a normal covalent boron-nitrogen bond in addition to a coordinated bond with a second nitrogen atom was first reported by GOUBEAU[1] as an intermediate in the thermal decomposition of the adduct of trimethyl-borane with ethylenediamine.

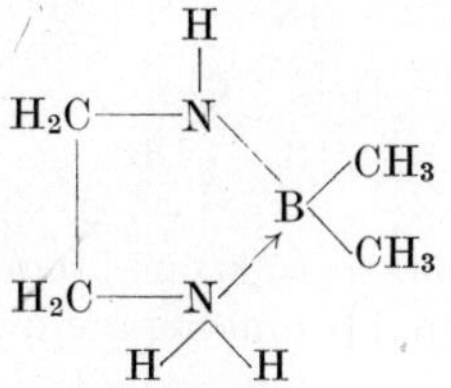

XXXVI

An analogous situation is exhibited by the recently described 2-bora-1,3-diazaazulenes[2]. The hydrolytic stability of the boron-nitrogen linkage

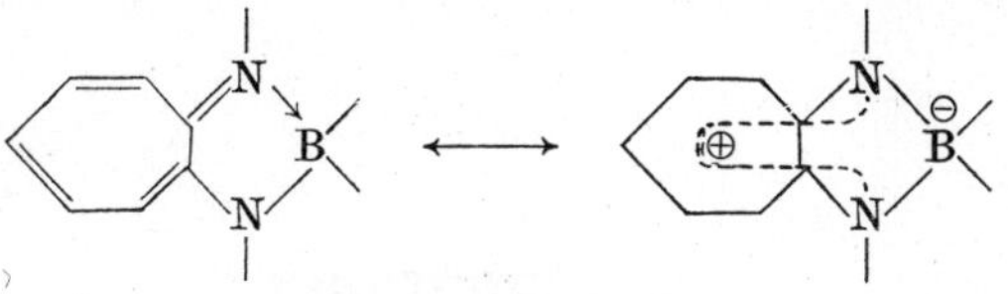

(a)   XXXVII   (b)

in this system indicates an appreciable electron delocalization. This event is substantiated by proton magnetic resonance studies, justifying the formulation of the indicated resonance structure, XXXVIIb.

HESSE and WITTE[3] studied the reaction of trialkylboranes with phenylisonitrile, C_6H_5NC. At room temperature, a novel cyclic system, 2,5-diboradihydropyrazine, XXXVIII, was obtained. At temperatures above 200° rearrangement occurs with the formation of a 2,5-dibora-

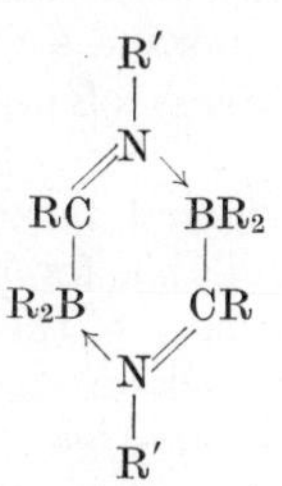

$R' = C_6H_5$
$R = C_2H_5$, m.p. 144°
$R = n\text{-}C_4H_9$, m.p. 125°

XXXVIII

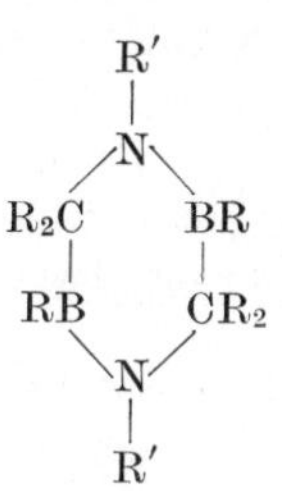

$R' = C_6H_5$

XXXIX

piperazine, XXXIX. The infrared spectra of the piperazine analogs show strong absorption in the 1400 cm.$^{-1}$ region, indicating a boron-nitrogen linkage with double bond character. The reaction of biguanidine

[1] GOUBEAU, J., and A. ZAPPEL: Z. anorg. allg. Chem. **279**, 38 (1955).
[2] HOLMQUIST, H. E., and R. E. BENSON: J. Amer. chem. Soc. **84**, 4720 (1962).
[3] HESSE, G., and H. WITTE: Angew. Chem. **75**, 791 (1963).

with either aminoboranes, R_2N—BR_2, or hydroxyboranes, $HOBR_2$, yields the new borinide ring system, XL. The borinides do not show the B—N

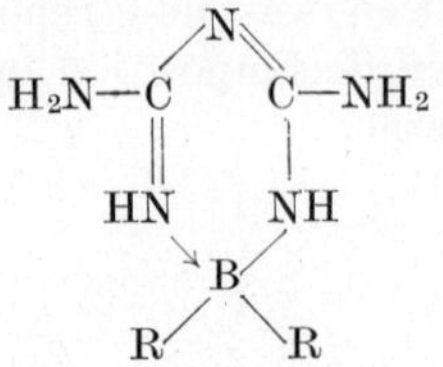

$$H_2N—C \quad C—NH_2$$

HN   NH

B

R   R

XL

absorption band in the 1350 to 1500 cm^{-1} region of the infrared spectrum which is typical for a normal aminoborane linkage. In conjunction with some [11]B nuclear magnetic resonance data, it was concluded that boron in these compounds is tercovalent in analogy to the tetraphenylborate ion, $[B(C_6H_5)_4]^{\ominus}$ [1].

Chapter II

Aminoboranes

A. The Nature of the B—N Bond in Aminoboranes

On the basis of certain physical constants of aminoboranes, $\begin{smallmatrix} R \\ R \end{smallmatrix} N—B \begin{smallmatrix} R \\ R \end{smallmatrix}$, which are consistently, numerically similar to those of analogous alkenes, WIBERG[2] has postulated the existence of a variable degree of B—N double bond character for the aminoborane system. This structure results from the participation of the nitrogen's lone-pair of electrons in the B—N bond.

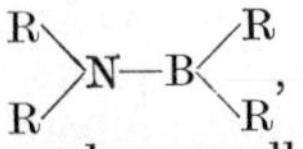

The concept of double bonding recognizes that trigonal boron in the ground state has two 2 s and one 2 p electrons available for bond formation. Therefore, in the basic aminoborane formula (I), the trivalent boron does not have a complete shell of valence electrons and tends to stabilize itself by back-coordination of the unshared electrons of the nitrogen to the boron (II). The planarity or near-planarity of the three covalent bonds of boron is a prerequisite for double bond character as has been demonstrated by GOUBEAU and BECHER for the aminoborane system[3, 4]. In addition, the force constants of some

---

[1] MILKS, J. E., G. W. KENNERLY and J. H. POLEVY: J. Amer. chem. Soc. 84, 2529 (1962).

[2] WIBERG, E.: Naturwissenschaften 35, 182, 212 (1948).

[3] GOUBEAU, J., and H. J. BECHER: Z. anorg. allg. Chem. 268, 133 (1952).

[4] GOUBEAU, J., M. RAHTZ and H. J. BECHER: ibid. 275, 161 (1954).

aminoboranes[1] indicate that the B—N bond order at room temperature is about 1.8, evidencing the probability of boron-nitrogen multiple bonding. Therefore, resonance structures as formulated by WIBERG (I, II) seem to be well founded.

Structure II implies charge transfer from nitrogen to boron; however, BECHER[2], in studying some methylated aminoboranes, found the bond moments to be practically zero. Apparently the dipole moment expected from formula II is considerably reduced by an asymmetrical electron sharing (in the sense B → N) owing to the electronegativity difference between boron and nitrogen. Therefore, BECHER concluded that formulation of a third structure, involving negatively charged nitrogen and positive boron, would be necessary in order to more accurately describe the bonding situation in aminoboranes. Analogous conclusions have been reached by COATES and LIVINGSTONE[3]. Indeed, molecular orbital calculations[4] by an extended HÜCKEL LCAO—MO method clearly illustrate that, in aminoboranes, the nitrogen bears a larger negative *net* charge than the boron. Population analysis demonstrates that, in the π-system, 0.23 electrons are transferred from nitrogen to boron, but the effect in the σ-system is reversed; the total charge transfer amounts to 0.28 electrons from boron to nitrogen. Naturally, simple molecular orbital calculations do not give an absolutely accurate picture although they are qualitatively correct. However, several conclusions can be reached on the basis of this theoretical treatment. For instance, calculations indicate the existence of a barrier to internal rotation in the aminoborane system in the order of about 10 kcal/mole. The probability of this occurence was disclosed in 1960 by NIEDENZU and DAWSON[5] based on experimental observations. Nuclear magnetic resonance investigations confirmed the concept of restricted rotation somewhat later[6-10] but it has been demonstrated that n.m.r. data alone do not suffice as unambigous proof[11].

[1] GOUBEAU, J.: Advances in Chemistry **42**, 87 (1964).
[2] BECHER, H. J.: Z. anorg. allg. Chem. **270**, 273 (1952).
[3] COATES, G. E., and J. G. Livingstone, J. chem. Soc. (London) **1961**, 1000.
[4] HOFFMANN, R.: Advances in Chemistry **42**, 78 (1964).
[5] NIEDENZU, K., and J. W. DAWSON: J. Amer. chem. Soc. **82**, 4223 (1960).
[6] RYSCHKEWITSCH, G. E., W. S. BREY and A. SAJI: ibid. **83**, 1010 (1961).
[7] BARFIELD, P. A., M. F. LAPPERT and J. LEE: Proc. chem. Soc. (London) **1961**, 421.
[8] BREY, W. S., M. E. FULLER II, G. E. RYSCHKEWITSCH and A. S. MARSHALL: Advances in Chemistry **42**, 100 (1964).
[9] TOTANI, T., H. WATANABE, T. NAKAGAWA, O. OHASHI and M. KUBO: ibid **42**, 108 (1964).
[10] NIEDENZU, K., H. BEYER, J. W. DAWSON and H. JENNE: Chem. Ber. **96**, 2653 (1963).
[11] BAECHLE, H., H. J. BECHER, H. BEYER, W. S. BREY, J. W. DAWSON, M. E. FULLER II and K. NIEDENZU: Inorg. Chem. **2**, 1065 (1963).

Nuclear magnetic resonance studies such as the aforementioned are based on the following: In an unsymmetrically substituted aminoborane, of the type $(CH_3)_2B—NRR'$(III), the two methyl groups are magnetically non-equivalent if hindered rotation about the B—N linkage exists.

$$H_3C \diagdown \qquad \diagup R$$
$$B \leftarrow N$$
$$H_3C \diagup \qquad \diagdown R'$$
III

Therefore, the ^{1}H resonance spectrum will indicate two methyl signals with different chemical shifts. At higher temperatures the rotation about the B → N bond will increase until only evidence of free rotation can be found: The two methyl groups become equivalent, and their n.m.r. signals combine. The energy of activation of rotation can be calculated from the temperature dependency of the chemical shift. Thus far, experimental values in the order of 10—15 kcal./mole have been found and are in good agreement with the theoretical value.

The relatively low values found for the rotational barrier of aminoboranes (as compared to those of alkenes) decrease the probability of isolating cis and trans isomers although the B—N bond order of aminoboranes is greatly influenced by substituents on the basic =N—B= grouping. Again, this was confirmed by molecular orbital calculations[1], but had been studied earlier by various investigators by means of infrared spectroscopy[2-8]. In a somewhat simplified approach, the position of the unsymmetrical B—N stretching frequency serves as an indicator of the bond order, wherein lower frequencies reflect the existence of lower bond orders. The general range for $\nu_{BN, asymm.}$ of aminoboranes was found from about 1350 to 1500 cm.$^{-1}$ [3] and spectral shifts were related to the effects of substitution at the aminoborane nucleus.

Table II-1. *B—N Stretching Bands of Aminoboranes, $RR'B—NR''R'''$*

R	R'	R''	R'''	ν_{BN}, cm.$^{-1}$
phenyl	phenyl	p-tolyl	p-tolyl	1361
phenyl	phenyl	phenyl	phenyl	1372
phenyl	methyl	phenyl	methyl	1381
phenyl	ethyl	phenyl	ethyl	1383
p-tolyl	p-tolyl	phenyl	phenyl	1385
methyl	methyl	methyl	phenyl	1388
p-tolyl	p-tolyl	methyl	methyl	1410
phenyl	ethyl	methyl	n-butyl	1412
phenyl	methyl	methyl	methyl	1417
ethyl	ethyl	ethyl	ethyl	1490
methyl	methyl	methyl	methyl	1530

[1] HOFFMAN, R.: Advances in Chemistry **42**, 78 (1964).
[2] GOUBEAU, J., and H. J. BECHER: Z. anorg. allg. Chem. **268**, 133 (1952).
[3] NIEDENZU, K., and J. W. DAWSON: J. Amer. chem. Soc. **81**, 5553 (1959).
[4] WYMAN, G. M., K. NIEDENZU and J. W. DAWSON: J. chem. Soc. (London) **1962**, 4068.
[5] GOUBEAU, J., M. RAHTZ and H. J. BECHER: Z. anorg. allg. Chem. **275**, 161 (1954).
[6] BECHER, H. J.: ibid. **289**, 262 (1957).
[7] BECHER, H. J.: Spectrochim. Acta **19**, 575 1963.
[8] BECHER, H. J.: Advances in Chemistry **42**, 71 (1964).

For instance, known examples of N-arylation show a decrease in the B—N bond order as compared with that found for N-alkylated aminoboranes. This is evidenced by a shift of the B—N frequency to lower values and is due to the resonance of the nitrogen's lone pair of electrons with the aromatic ring. A transitory N—C$_{aryl}$ double bond may be formed and the electron density at the boron atom decreases correspondingly. Aryl groups at the boron show a smaller effect, although, even in this case, resonance interaction of the π-electrons of the aromatic system with the electron deficient boron is possible. Thus the electron density of the boron is not diminished on an absolute basis, although the B—N bond order is lessened[1]. On replacement of N-attached hydrogen by deuterium or alkyl groups, ν_{BN} shifts to higher frequencies indicating a higher B—N bond order and the same relation holds true for the replacement of B-alkyl groups by halogen[2]. The experimental results appear to be in agreement with molecular orbital calculations[3, 4].

Summarizing, it can be said that experimental and theoretical data substantiate the concept of variable double bond character in the aminoborane system. However, it should be recognized that the magnitude of this character is by no means comparable to that of corresponding alkenes. This is further exemplified by the observation that although the ultraviolet spectra of (diphenylamino)diphenylborane, $(C_6H_5)_2N$—$B(C_6H_5)_2$, and tetraphenylethylene are very similar in general appearance[5], their X-ray diffraction patterns clearly indicate that these two isoelectronic compounds are not isomorphous[6]. On the other hand, the interconvertible *cis-trans* isomerism in the aminoborane system has most recently been verified by complete assignments of proton magnetic resonance spectra of a series of unsymmetric aminoboranes and also by infrared spectroscopy[7].

The charge distribution in aminoboranes as illustrated in formula IVa is misleading. IVb is a better picture of the actual *total* charge distribution on the basis of theoretical data, but is in lesser conformance with experiment. It therefore appears reasonable to illustrate aminoboranes conventionally by formula IVc, i.e. a double-bonded molecule without any assignment of charges.

$$\underset{IV\,a}{\diagdown \underset{\ominus}{B}\!\!=\!\!\underset{\oplus}{N}\diagup} \qquad \underset{IV\,b}{\diagdown \underset{\oplus}{B}\!\!=\!\!\underset{\ominus}{N}\diagup} \qquad \underset{IV\,c}{\diagdown B\!\leftarrow\!N\diagup}$$

The disadvantage of formula IVc resides in the fact that the indicated structure implies double bonding on a quantitative level rather than pri-

---

[1] WYMAN, G. M., K. NIEDENZU and J. W. DAWSON: J. chem. Soc. (London) 1962, 4068.

[2] BECHER, H. J.: Spectrochim. Acta **19**, 575 (1963).

[3] HOFFMANN, R.: Advances in Chemistry **42**, 78 (1964).

[4] KAUFMAN, J. J., and J. R. HAMANN: ibid. **42**, 95 (1964).

[5] COATES, G. E., and J. G. LIVINGSTONE: J. chem. Soc. (London) **1961**, 1000.

[6] NIEDENZU, K., H. BEYER and J. W. DAWSON: Inorg. Chem. **1**, 738 (1962).

[7] KUBO, M.: personal communication.

marily qualitative. Hence, for all practical purposes, it is best to use
IVd with the understanding that the unshared electron pair of the
nitrogen can participate in the boron-nitrogen linkage.

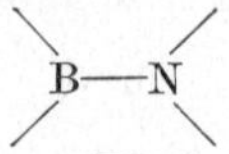

$$\diagdown B\!\!-\!\!N\diagdown$$

IVd

B. Monoaminoboranes

1. General Remarks

The parent compound of the aminoborane system, $H_2N\!-\!BH_2$, is not
known in monomeric form. It has long been thought to be highly polymeric
and unstable[1] but more recent investigations indicate that such may
not be the case (see Chapter II-C). The extent of association of aminoborane decreases as hydrogen is replaced by organic groups. Thus
methylaminoborane, $CH_3HN\!-\!BH_2$, exists as a stable trimer[2] and
reversible monomer-dimer equilibria have been demonstrated for dimethylaminoborane[3], $(CH_3)_2N\!-\!BH_2$, (dimethylamino)methylborane[4],
$(CH_3)_2N\!-\!BHCH_3$, (methylamino)dimethylborane[5], $CH_3HN\!-\!B(CH_3)_2$,
aminodimethylborane[6], $H_2N\!-\!B(CH_3)_2$, aminodiphenylborane[7], $H_2N\!-\!B$
$(C_6H_5)_2$, and (methylamino)diphenylborane[6], $CH_3HN\!-\!B(C_6H_5)_2$. The
rapid and spontaneous dimerization of (dimethylamino)dichloroborane,
$(CH_3)_2N\!-\!BCl_2$, was the first self-association of an aminoborane to be
recognized[8]. The low order of its chemical reactivity led to proposing
the existence of a cyclobutane structure for the dimer (V). This concept
is substantiated by the small dipole moment of the compound. Additional

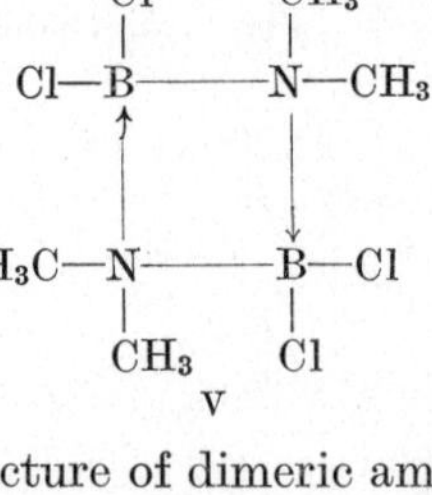

support for the cyclic structure of dimeric aminoboranes is provided by
infrared spectroscopy, since the B—N stretching frequency of such

[1] SCHAEFFER, G. W., M. D. ADAMS and F. J. KOENIG: J. Amer. chem. Soc.
78, 725 (1956).
[2] BISSOT, T. C., and R. W. PARRY: ibid. 77, 3481 (1955).
[3] WIBERG, E., A. BOLZ and P. BUCHHEIT: Z. anorg. allg. Chem. 256, 285 (1948).
[4] BURG, A. B., and J. L. BOONE: J. Amer. chem. Soc. 78, 1521 (1956).
[5] WIBERG, E., and K. HERTWIG: Z. anorg. allg. Chem. 255, 141 (1947).
[6] BUCHHEIT, P.: Dissertation, University of Munich, Germany, 1942.
[7] COATES, G. E., and J. G. LIVINGSTONE: J. chem. Soc. (London) 1961, 1000.
[8] WIBERG, E., and K. SCHUSTER: Z. anorg. allg. Chem. 213, 77, 89 (1933).

molecules is usually observed in the 900 cm.$^{-1}$ region, thus presumably excluding participation of the free electron pair of the nitrogen in a B—N double bond. Although a variety of aminodihalogenoboranes, $R_2N—BX_2$, dimerize on standing to cyclic structures of type V[1], (diethylamino)dichloroborane, $(C_2H_5)_2N—BCl_2$, does not. Therefore, it appears likely that steric factors may often control the dimerization of aminoboranes. For example, the small steric requirements of N-substituents in (dimethylamino)dichloroborane, piperidinodichloroborane or morpholinodichloroborane permit intermolecular coordinate bond formation and consequent cyclization. The bulkier diethylamino or di-n-butylamino groups do not allow cyclization to occur and (diethylamino)-dichloroborane and (di-n-butylamino)dichloroborane are known only as monomers. In consonance with these findings, the much slower dimerization of (dimethylamino)dibromoborane as compared to that of (dimethylamino)dichloroborane can be attributed to the increased spatial requirements of the BBr_2 group.

Summarizing, dimeric species are known only for those aminoboranes having hydrogen and/or halogen as one or more of the substituents on the basic B—N grouping. Dimerization seems to depend on steric factors and can be reversed by distillation[1] or sublimation[2]. Tetraorgano-substitued aminoboranes, $R_2N—BR_2$, are known only in the monomeric form, although slight association has been reported for some concentrated solutions[3]. Conventional molecular weight determinations are not always unequivocal. However, ^{11}B nuclear magnetic resonance usually permits one to estimate the grade of association[4-6] (see Appendix).

BURG and KULJIAN have reported the formation of a highly polymeric aminoborane by heating disilylaminoborane[7], $(H_3Si)_2N—BH_2$); the synthesis of an isopropylaminoborane polymer accompanied by the evolution of hydrogen is alleged to result from heating the parent $(C_3H_7)HN—BH_2$ [8]. Only one example of converting a low molecular weight aminoborane to a high degree of polymerization appears to be established. When dimeric dimethylaminoborane is subjected to 3000 atmospheres pressure at 150°, an amorphous, insoluble and infusible polymer is obtained. It appears to be stable towards hydrolysis but decomposes at high temperatures. After standing at room temperature for a few months, the material reverts almost completely to the dimer[9].

[1] MUSGRAVE, O. C.: J. chem. Soc. (London) **1956**, 4305.
[2] BROWN, J. F.: J. Amer. chem. Soc. **74**, 1219 (1951).
[3] COATES, G. E., and J. G. LIVINGSTONE: J. chem. Soc. (London) **1961**, 1000.
[4] PHILLIPS, W. D., H. C. MILLER and E. L. MUETTERTIES: J. Amer. chem. Soc. **81**, 4496 (1959).
[5] HAWTHORNE, M. F.: ibid. **83**, 2671 (1961).
[6] RUFF, J. K.: J. org. Chemistry **27**, 1020 (1962).
[7] BURG, A. B., and E. S. KULJIAN: J. Amer. chem. Soc. **72**, 3103 (1950).
[8] HOUGH, W. V., and G. W. SCHAEFFER: U.S. Patent 2,809,171 (1957).
[9] DEWING, I.: Angew. Chem. **73**, 681 (1961).

2. Organic Substituted Aminoboranes

Earlier preparations of organic substituted aminoboranes were influenced by the fact that amine-boranes decompose at elevated temperatures. Thus dimethylamine-borane loses hydrogen at 200° to afford dimethylaminoborane[1, 2].

$$(CH_3)_2HN \cdot BH_3 \xrightarrow{200°} (CH_3)_2N{-}BH_2 + H_2 \qquad \text{(II-1}$$

Similarly, ammonia-trimethylborane eliminates methane at 280° and 20 atmospheres pressure yielding aminodimethylborane, $H_2N{-}B(CH_3)_2$ [3]. The same compound is obtained from the thermal decomposition of a mixture of ammonia and methyldiborane[4]. (Dimethylamino)dimethylborane is likewise formed through pyrolysis of dimethylamine-trimethylborane. The preparation of organic substituted aminoboranes can be illustrated by the general equation:

$$R_3N \cdot BR_3' \xrightarrow{\Delta} R_2N{-}BR_2' + R{-}R' \qquad \text{(II-2)}$$

A wide variety of boron hydride derivatives can be utilized in this reaction[5]. However, this synthetic route is rendered awkward by the necessity of operating in high vacuum systems. A considerable improvement was effected by SCHAEFFER and ANDERSON[6], who reported the preparation of aminoboranes through the reaction of metal hydroborates with amine salts.

$$R_2NH_2Cl + LiBH_4 \rightarrow 2H_2 + LiCl + R_2N{-}BH_2 \qquad \text{(II-3)}$$

The GRIGNARD reaction of aminodihalogenoboranes, $R_2N{-}BX_2$ [7, 8] which can be effected with ordinary laboratory equipment appears to be even more convenient. Since aminodihalogenoboranes are readily obtained by dehydrohalogenation of amine-trihalogenoboranes with tertiary amines[9], this method provides easy access to a variety of aminoboranes.

$$R_2HN \cdot BX_3 + R_3'N \rightarrow R_2N{-}BX_2 + [R_3'NH]X \qquad \text{(II-4)}$$

$$R_2N{-}BX_2 + 2R''MgX \rightarrow R_2N{-}BR_2'' + 2MgX_2 \qquad \text{(II-5)}$$

Aminoboranes with completely different substituents on the boron and nitrogen can be synthesized by reacting a properly substituted

[1] WIBERG, E., A. BOLZ and P. BUCHHEIT: Z. anorg. allg. Chem. 256, 285 (1948).
[2] BURG, A. B., and C. L. RANDOLPH: J. Amer. chem. Soc. 71, 3451 (1949).
[3] WIBERG, E., K. HERTWIG and A. BOLZ: Z. anorg. allg. Chem. 265, 177 (1948).
[4] SCHLESINGER, H. I., L. HORVITZ and A. B. BURG: J. Amer. chem. Soc. 58, 409 (1936).
[5] HAWTHORNE, M. F.: ibid. 83, 2671 (1961).
[6] SCHAEFFER, G. W., and E. R. ANDERSON: ibid. 71, 2143 (1949).
[7] NIEDENZU, K., and J. W. DAWSON: ibid. 81, 5553 (1959).
[8] NIEDENZU, K., and J. W. DAWSON: ibid. 82, 4223 (1960).
[9] BROWN, J. F.: ibid. 74, 1219 (1951).

aminomonohalogenoborane with the appropriate GRIGNARD reagent[1].
Such derivatives, VI, have played an important part in the elucidation of
the nature of the boron-nitrogen bond in the aminoborane system partic-

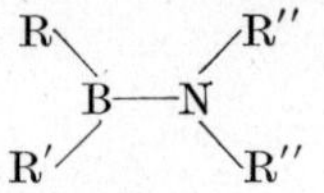

VI

ularly through study of their structure by nuclear magnetic resonance
spectroscopy.

Other preparative approaches to organic substituted aminoboranes
involve the disproportionation of trisaminoboranes with triorganoboranes[2]

$$B(NR_2)_3 + 2 BR_3' \rightarrow 3 R_2'B{-}NR_2 \tag{II-6}$$

and transamination reactions

$$R_2N{-}BR_2 + HNR_2' \rightleftarrows R_2'N{-}BR_2 + R_2NH \tag{II-7}$$

This last equilibrium reaction has recently been under intensive study
in various laboratories[3-6]. Its application in boron chemistry was first
reported by GERRARD and coworkers[7]. The mild conditions associated
with transamination have provided access to a wide variety of boron-
nitrogen compounds, which are unstable under more stringent conditions
of preparation. Trans*boronation* has been achieved in an analogous
manner. For example, aminoboranes readily react with alkylboranes
or alkyldiboranes as illustrated by the equation:

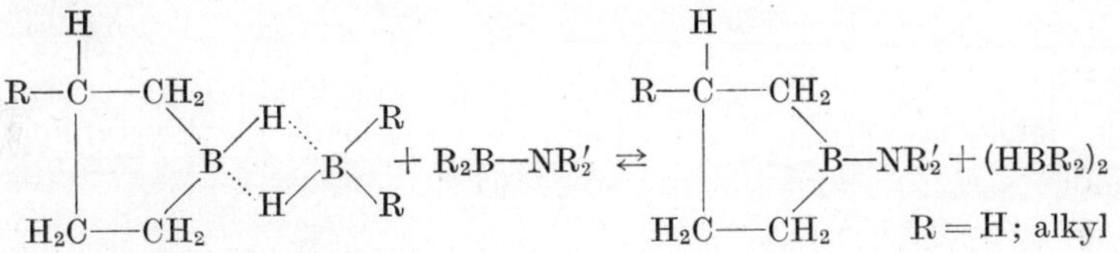

Aminoboranes, with the boron incorporated in a carbon ring, are
preferentially formed in this equilibrium reaction[8]. Such materials have
also been obtained on treatment of dimethylaminoborane, $(CH_3)_2N{-}BH_2$,
with butadiene at 150° and 3,000 atmospheres pressure[9].

The action of $Al[N(CH_3)_2]_3$ on various organoboron compounds (trialkylboranes, boroxines, etc.) has been found advantageous for the

[1] NIEDENZU, K., and J. W. DAWSON: J. Amer. chem. Soc. 82, 4223 (1960).
[2] NIEDENZU, K., H. BEYER, J. W. DAWSON and H. JENNE: Chem. Ber. 96,
2653 (1963).
[3] ENGLISH, W. D., A. L. McCLOSKEY and H. STEINBERG: J. Amer. chem. Soc.
83, 2122 (1961).
[4] NIEDENZU, K., D. H. HARRELSON and J. W. DAWSON: Chem. Ber. 94, 671 (1961).
[5] NÖTH, H.: Z. Naturforsch. 16b, 470 (1961).
[6] NIEDENZU, K., H. BEYER and J. W. DAWSON: Inorg. Chem. 1, 738 (1962).
[7] GERRARD, W., M. F. LAPPERT and C. A. PEARCE: J. chem. Soc. (London)
1957, 381.
[8] KÖSTER, R., and K. IWASAKI: Advances in Chemistry 42, 148 (1964).
[9] DEWING, I.: Angew. Chem. 73, 681 (1961).

preparation of dimethylaminoboranes[1]. However, aminolysis of di-organohalogenoboranes (eq. II-9) has not yet been fully explored. This

$$\begin{matrix} R \\ \\ R \end{matrix}\!\!>\!B\!-\!X + 2\,HN\!\!<\!\begin{matrix} R' \\ \\ R' \end{matrix} \rightarrow \begin{matrix} R \\ \\ R \end{matrix}\!\!>\!B\!-\!N\!\!<\!\begin{matrix} R' \\ \\ R' \end{matrix} + [R_2'NH_2]X \qquad (\text{II-9})$$

situation could be explained by the fact that compounds of the type R_2BX were hard to prepare or obtain. Since McCusker and coworkers have studied the disproportionation of trihalogenoboranes with tri-organoboranes[2, 3], however, R_2BX compounds are more easily available and have been used effectively for the preparation of aminoboranes by simple aminolysis[4].

A number of N-silylated aminoboranes have recently been described. They are readily obtained through transamination of an aminoborane with a silylamine[5]; another approach to the preparation of such materials is based upon earlier observations[6-9] that a silicon-nitrogen linkage is readily cleaved when attacked by a halogenoborane[5, 10].

$$\begin{matrix} R & H \\ | & | \\ R\!-\!B & + & |N\!-\!Si(CH_3)_3 \\ | & | \\ X & Si(CH_3)_3 \end{matrix} \longrightarrow \begin{matrix} R & H \\ | & | \\ R\!-\!B\!\leftarrow\!N\!-\!Si(CH_3)_3 \\ | & | \\ X & Si(CH_3)_3 \end{matrix}$$

$$\longrightarrow \begin{matrix} R \\ \\ R \end{matrix}\!\!>\!B\!-\!N\!\!<\!\begin{matrix} H \\ \\ Si(CH_3)_3 \end{matrix} + X\!-\!Si(CH_3)_3 \qquad (\text{II-10})$$

Some typical representatives of substituted aminoboranes are found in Table II-2.

Steric effects appear to have a pronounced influence on the stability of aminoboranes. Most of these compounds are very sensitive towards hydrolysis. Indeed, N-phenylated compounds seem to be especially unstable towards any kind of chemical attack and it is noteworthy that, with a decreasing double bond character of the B—N linkage, the stability towards water, alcohols etc. decreases[11]. No detailed investigations have been reported, but it can be assumed that, by decreasing the B—N bond order, the possibility for a Lewis acid—Lewis base

[1] Ruff, J. K.: J. org. Chemistry 27, 1020 (1962).
[2] McCusker, P. A., G. F. Hennion and E. C. Ashby: J. Amer. chem. Soc. 79, 5192 (1957).
[3] McCusker, P. A., and L. J. Glunz: ibid. 77, 4253 (1955).
[4] Niedenzu, K., D. H. Harrelson, W. George and J. W. Dawson: J. org. Chemistry 26, 3037 (1961).
[5] Jenne, H., and K. Niedenzu: Inorg. Chem. 3, 68 (1964).
[6] Burg, A. B., and E. S. Kuljian: J. Amer. chem. Soc. 72, 3103 (1950).
[7] Sujishi, S., and S. Witz: ibid. 79, 2447 (1957).
[8] Ebsworth, E. A. V., and H. J. Emeleus: J. chem. Soc. (London) 1958, 2150.
[9] Becke-Goehring, M., and H. Krill: Chem. Ber. 94, 1059 (1961).
[10] Nöth, H.: Z. Naturforsch. 16b, 618 (1961).
[11] Wyman, G. M., K. Niedenzu and J. W. Dawson: J. chem. Soc. (London) 1962, 4068.

Table II-2. *Organic Substituted Aminoboranes,* $\begin{array}{c} R \\ R' \end{array}\!\!>\!\!B\!-\!N\!<\!\!\begin{array}{c} R'' \\ R''' \end{array}$

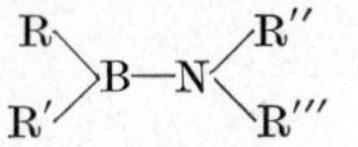

R	R'	R''	R'''	b.p. (mm.)°C	Ref.
methyl	methyl	methyl	methyl	49—51	2
n-propyl	n-propyl	methyl	methyl	42 (9)	8
methyl	methyl	butyl	hydrogen	114 (722)	7
n-butyl	n-butyl	hydrogen	hydrogen	64—66 (1.7)	4
t-butyl	hydrogen	ethyl	ethyl	68 (46)	3
methyl	methyl	ethyl	phenyl	51 (4)	5
ethyl	ethyl	ethyl	phenyl	158—165 (6)	5
methyl	methyl	methyl	α-naphthyl	89 (3)	5
phenyl	phenyl	methyl	methyl	102—104 (0.05)	1
phenyl	phenyl	ethyl	ethyl	161 (11)	5
methyl	phenyl	methyl	phenyl	98 (3)	6
n-hexyl	phenyl	methyl	cyclohexyl	102—104 (3)	4
benzyl	phenyl	benzyl	methyl	170—180 (3)	4

interaction is increased. It has been shown[9, 10] that the addition of acceptor molecules is directly related to their acceptor power: Trimethyl-borane, $B(CH_3)_3$, and dimethylfluoroborane, $(CH_3)_2BF$, do not react with aminoboranes at low temperatures. Trichloroborane and trifluoroborane however, form addition compounds which decompose at even low

$$(CH_3)_2N\!-\!B(CH_3)_2 \cdot BX_3 \rightarrow (CH_3)_2BX + (CH_3)_2N\!-\!BX_2 \qquad (II\text{-}11)$$

temperatures. This disproportionation reaction can be interpreted in terms of a cyclic intermediate (VII) formed in a four center reaction.

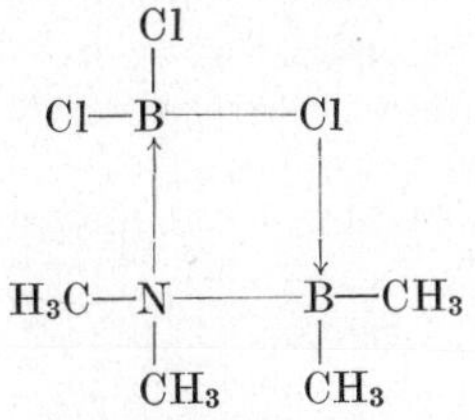

$$
\begin{array}{ccc}
 & Cl & \\
 & | & \\
Cl\!-\!B\!\!-\!\!\!-\!\!\!-\!Cl & \\
 & \uparrow \qquad | & \\
H_3C\!-\!N\!\!-\!\!\!-\!\!\!-\!B\!-\!CH_3 & \\
 & | \qquad | & \\
 & CH_3 \quad CH_3 & \\
 & VII &
\end{array}
$$

1 COATES, G. E., and J. G. LIVINGSTONE: J. chem. Soc. (London) **1961,** 1000.
2 ERICKSON, C. E., and F. C. GUNDERLOY: J. org. Chemistry **24,** 1161 (1959).
3 HAWTHORNE, M. F.: J. Amer. chem. Soc. **83,** 2671 (1961).
4 NIEDENZU, K., H. BEYER, J. W. DAWSON and H. JENNE: Chem. Ber. **96,** 2653 (1963).
5 NIEDENZU, K., and J. W. DAWSON: J. Amer. chem. Soc. **81,** 5553 (1959).
6 NIEDENZU, K., and J. W. DAWSON: ibid. **82,** 4223 (1960).
7 NÖTH, H.: Z. Naturforsch. **16 b,** 470 (1961).
8 RUFF, J. K.: J. org. Chemistry **27,** 1020 (1962).
9 BURG, A. B., and J. BANUS: J. Amer. chem. Soc. **76,** 3903 (1954).
10 BECHER, H. J.: Z. anorg. allg. Chem. **288,** 235 (1955).

Preparative

a. (Diethylamino)diethylborane by a Grignard Reaction[1]

To a solution of 30.8 g. (0.2 mole) of (diethylamino)dichloroborane in 400 cc. of dry ether, 12 g. (0.5 mole) of magnesium are added. The mixture is stirred and a solution of 45.8 g. (0.42 mole) of ethyl bromide in 200 cc. of dry ether is added at such a rate that the reaction mixture refluxes steadily. When addition is complete, the reaction is refluxed for an additional three hours. After the reaction mixture is filtered, the ether is removed from the filtrate and the residue is distilled at normal pressure. Yield: 22 g. (78%) of (diethylamino)diethylborane, b.p. 154°.

b. (Di-n-pentylamino)di-n-pentylborane by Aminolysis[2]

A solution of 22.1 g. (0.114 mole) of di-n-pentylchloroborane in 100 cc. of dry benzene is added with stirring to a solution of 18.5 g. (0.117 mole) of di-n-pentyl-amine in 500 cc. of benzene. The resultant exothermic reaction produces a white precipitate, which dissolves on refluxing with the release of hydrogen chloride. The mixture is refluxed for several hours until the evolution of hydrogen chloride has almost ceased; after filtration, the solvent is removed and the residue is distilled in vacuum affording 24 g. (69%) of (di-n-pentylamino)di-n-pentylborane, b.p. 124° (8 mm.).

c. (Diphenylamino)diphenylborane by Aminolysis (alternate method)[3]

Diphenylamine in benzene is added to a solution containing equivalent amounts of triethylamine and diphenylchloroborane in the same solvent. The mixture is refluxed for thirty minutes and filtered under nitrogen from the precipitated triethylammonium chloride. The filtrate is concentrated to crystallization to afford a yield of 61% of (diphenylamino)diphenylborane, m.p. 255—256°.

3. Aminohalogenoboranes

Aminodihalogenoboranes, $R_2N—BX_2$, were first prepared more than sixty years ago by MICHAELIS through the interaction of trihalogeno-boranes (BCl_3, BBr_3) and secondary amines[4]. It was shown later that this reaction occurs step-wise and that the nature of the products depends upon the relative proportions of the reactants[5-7]. However, yields of aminodihalogenoboranes are normally poor[8] even though an organic diluent be used, since the resultant hydrogen halide can react with both, the aminohalogenoborane and the amine[9, 10]. This basic preparative procedure is considerably improved by running the reaction in the presence of triethylamine[11] (eq. II-12).

$$R_2NH + BX_3 \rightarrow R_2HN \cdot BX_3 \xrightarrow{N(C_2H_5)_3} R_2N—BX_2 + [(C_2H_5)_3NH]X \quad (II\text{-}12)$$

[1] NIEDENZU, K., and J. W. DAWSON: J. Amer. chem. Soc. **82**, 4223 (1960).
[2] NIEDENZU, K., D. H. HARRELSON, W. GEORGE and J. W. DAWSON: J. org. Chemistry **26**, 3037 (1961).
[3] COATES, G. E., and J. G. LIVINGSTONE: J. chem. Soc. (London) **1961**, 1000.
[4] MICHAELIS, A., and K. LUXEMBOURG: Ber. dtsch. chem. Ges. **29**, 710 (1896).
[5] WIBERG, E., and W. SÜTTERLIN: Z. anorg. allg. Chem. **202**, 46 (1931).
[6] WIBERG, E., and K. SCHUSTER: ibid. **213**, 77, 89 (1933).
[7] WIBERG, E., A. BOLZ and P. BUCHHEIT: ibid. **256**, 285 (1948).
[8] BROWN, C. A., and R. C. OSTHOFF: J. Amer. chem. Soc. **74**, 2340 (1952).
[9] GERRARD, W., M. F. LAPPERT and C. A. PEARCE: J. chem. Soc. (London) **1957**, 381.
[10] NÖTH, H., and S. LUKAS: Chem. Ber. **95**, 1505 (1962).
[11] BROWN, J. F.: J. Amer. chem. Soc. **74**, 1219 (1951).

Using this technique, reasonable yields have been reported[1, 2]. However, pyridine is not suitable as the HX acceptor in place of the triethylamine, since pyridine tends to promote the formation of borazylammonium salts (see Chapter I-C). The actual course of the dehydrohalogenation reaction has not yet been studied in detail. However, it seems reasonable to postulate a reaction sequence wherein a borazylammonium salt is

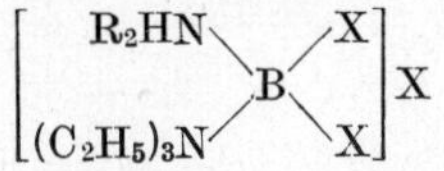

$$\begin{bmatrix} R_2HN & X \\ & B \\ (C_2H_5)_3N & X \end{bmatrix} X$$

formed as an intermediate unstable species which in turn is dehydrohalogenated by additional triethylamine. The resultant $R_2N-BX_2 \cdot N(C_2H_5)_3$ splits off triethylamine yielding the desired aminodihalogenoborane.

Aminodihalogenoboranes have also been obtained through the thermal decomposition of some amine-halogenoboranes[3], but this method is very inefficient. Most advantageous of all procedures appears to be the disproportionation of a trisaminoborane with a trihalogenoborane[4-7].

$$B(NR_2)_3 + 2 BX_3 \rightarrow 3 X_2B-NR_2 \qquad \text{(II-13)}$$

The action of hydrogen halides upon bisaminoboranes has been studied[5, 8, 9] and has been used to prepare aminomonohalogenoboranes. Aminohalogenoboranes of the type $RHN-BR'X$ were extremely rare prior to the development of this last synthesis. Only two examples had been described in the literature:

(Methylamino)methylchloroborane, $CH_3HN-BCl(CH_3)$, was obtained by the action of hydrogen chloride on B-trimethyl-N-trimethyl-borazine, $(-BCH_3-NCH_3-)_3$, at $150°$ [3]; by heating methylamine or dimethylamine with dimethylfluoroborane, (methylamino)methylfluoroborane, $(CH_3)HN-BF(CH_3)$, b.p. $46°$, and (dimethylamino)methylfluoroborane, $(CH_3)_2N-BF(CH_3)$, b.p. $58°$, respectively have been prepared[10].

Aminomonohalogenoorganoboranes, $R_2'N-BXR$ ($R' =$ alkyl, aryl), have been synthesized through the interaction of equimolar amounts of an organodihalogenoborane and a secondary amine in the presence of triethylamine[11]. They are also formed through the disproportionation

[1] NIEDENZU, K., and J. W. DAWSON: J. Amer. chem. Soc. **81**, 5553 (1959).
[2] MIKHAILOV, B. M., W. D. SCHTULJAKOW and T. A. SCHTSCHEGOLEWA: Proc. Acad. Sci. USSR. **1962**, 1698.
[3] WIBERG, E., and K. HERTWIG: Z. anorg. allg. Chem. **255**, 141 (1947).
[4] GERRARD, W., M. F. LAPPERT and C. A. PEARCE: J. chem. Soc. (London) **1957**, 381.
[5] NÖTH, H., and S. LUKAS: Chem. Ber. **95**, 1505 (1962).
[6] GOUBEAU, J., M. RAHTZ and H. J. BECHER: Z. anorg. allg. Chem. **275**, 161 (1954).
[7] BROTHERTON, R. J., A. L. MCCLOSKEY, L. L. PETTERSON and H. STEINBERG: J. Amer. chem. Soc. **82**, 6242 (1960).
[8] NÖTH, H, W. A. DOROCHOV, P. FRITZ and F. PFAB: Z. anorg. allg. Chem. **318**, 293 (1962).
[9] NÖTH, H., and P. FRITZ: ibid. **322**, 297 (1963).
[10] WIBERG, E., and G. HORELD: Z. Naturforsch. **6b**, 338 (1951).
[11] NIEDENZU, K., and J. W. DAWSON: J. Amer. chem. Soc. **82**, 4223 (1960).

of an aminoorganoborane with an aminohalogenoborane[1-3]. Both methods serve equally well for preparative purposes.

$$RBCl_2 + HNR_2' \xrightarrow[-HCl]{N(C_2H_5)_3} RB\begin{cases} NR_2' \\ Cl \end{cases} \qquad (II\text{-}14)$$

$$R_2N\text{—}BR_2' + R_2N\text{—}BX_2 \rightarrow 2R'B\begin{cases} NR_2 \\ X \end{cases} \qquad (II\text{-}15)$$

Monomeric aminohalogenoboranes hydrolyze quite readily. Also, the halogen can be replaced by hydrogen through the action of lithium-hydride[4]; replacement with organic groups is achieved through the action of metal-organic compounds as described above[5, 6].

$$R_2N\text{—}B\begin{cases} X' \\ R' \end{cases} + LiH \rightarrow R_2N\text{—}B\begin{cases} H \\ R' \end{cases} + LiX \qquad (II\text{-}16)$$

$$R_2N\text{—}B\begin{cases} X \\ R' \end{cases} + MeR'' \rightarrow R_2N\text{—}B\begin{cases} R'' \\ R' \end{cases} + MeX \qquad (II\text{-}17)$$

The heats of formation were determined for all members of the series $BCl_{3-n}[N(CH_3)_2]_n$ [7]. The data are illustrated in Table II-3 and

Table II-3. *Heats of Formation of the Series $BCl_{3-n}[N(CH_3)_2]_n$*

Compound	$\Delta Hf°$ (liq.)	ΔH vap.	$\Delta Hf°$ (g.)
BCl_3	−103.0	5.5	−97.5
$BCl_2N(CH_3)_2$	−108.0	8.9	−99.1
$BCl[N(CH_3)_2]_2$	− 95.25	10.0	−85.3
$B[N(CH_3)_2]_3$	− 77.1	11.2	−65.9

show, that the bond energy terms $D[B\text{—}Cl]$ and $D[B\text{—}N(CH_3)_2]$ are not constant within this series. For example, there is an overall enhancement in the bond-energy values by 12.1 kcal./mole in (dimethylamino)dichloroborane. In other words, this molecule is more stable by 12.1 kcal./mole than would be the case if the B—Cl bonds had the same mean energy as in trichloroborane. This difference can be explained in terms of the π-bonding which is superimposed on the trigonal σ-bonding, and the back-coordination from nitrogen to boron is more powerful than that from chlorine to boron.

[1] GUNDERLOY, F. C. and C. E. ERICKSON: Inorg. Chem. **1,** 349 (1962).
[2] NIEDENZU, K., H. BEYER, J. W. DAWSON and H. JENNE: Chem. Ber. **96,** 2653 (1963).
[3] NÖTH, H., and P. FRITZ: Z. anorg. allg. Chem. **324,** 270 (1963).
[4] NÖTH, H., W. A. DOROCHOV, P. FRITZ and F. PFAB: ibid. **318** 293 (1962).
[5] NIEDENZU, K., and J. W. DAWSON: J. Amer. chem. Soc. **81,** 5553 (1959).
[6] COATES, G. E.: J. chem. Soc. (London) **1950,** 3481.
[7] SKINNER, H. A., and N. B. SMITH: ibid. **1954,** 2324.

(Dimethylamino)dichloroborane

a. By Dehydrohalogenation of Dimethylamine-trichloroborane[1].

A solution of 41 ml. of trichloroborane in 200 ml. of benzene is added with stirring to a cooled solution of 33 ml. of dimethylamine in 300 ml. of benzene. After addition is complete the cooling bath is removed and the reaction mixture is stirred until it reaches room temperature. A solution of 70 ml. of triethylamine is then added with stirring and the mixture is refluxed for three hours. After standing overnight, it is filtered, the benzene is removed from the filtrate and the residue distilled. Yield: 45.5 g. (73%), b.p. 51—53° (90 mm).

b. By Disproportionation of Tris(dimethylamino)borane with Trichloroborane[2].

A solution of 143 g. tris(dimethylamino)borane in 250 ml. of dry hexane is cooled in an ice-bath and a cold solution of 164 ml. of trichloroborane in 200 ml. of hexane is added slowly with vigorous stirring. After warming the reaction mixture to room temperature, the solvent is stripped of the filtrate and the residue distilled under reduced pressure. Yield: About 90%.

4. Other Monoaminoborane Systems

Some other derivatives of the monoaminoborane system should be included here for the sake of completeness. Some information is available on aminodialkoxyboranes, R_2N—$B(OR')_2$, and (amino)alkoxyhalogenoboranes, R_2N—$BCl(OR')$ [3]. The amino groups in the former are very reactive. Alcohols or hydrogen halides attack the boron-nitrogen bond, and a transamination occurs when the molecules are subjected to reaction with amines. (Amino)alkoxyhalogenoboranes are obtained by partial aminolysis of alkoxydihalogenoboranes or through disproportionation of bis(dialkylamino)halogenoborane with bis(dialkoxy)halogenoborane; the products are easily hydrolyzed.

Some cyclic derivatives of similar nature (VIII) are also known[4]. The same is true for other ring systems with an exocyclic boron-nitrogen linkage. For example, the reaction of bisborolanes with secondary amines

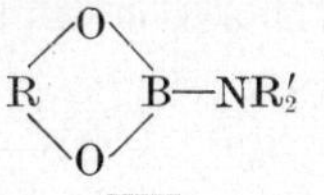

VIII

and subsequent dehydrogenation as illustrated in eq. II-17 has been described[5].

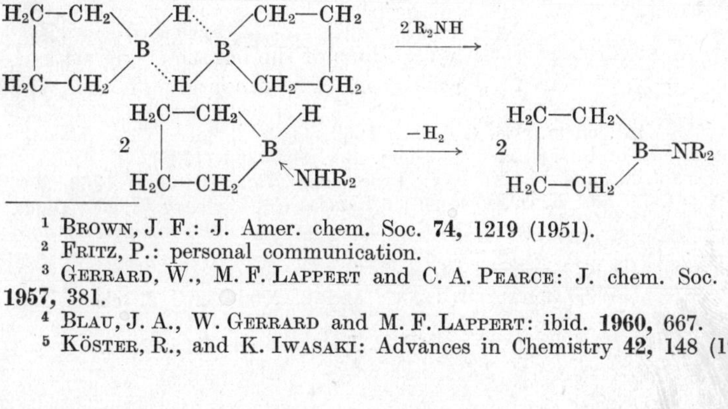

$$(II\text{-}18)$$

---

[1] BROWN, J. F.: J. Amer. chem. Soc. **74**, 1219 (1951).

[2] FRITZ, P.: personal communication.

[3] GERRARD, W., M. F. LAPPERT and C. A. PEARCE: J. chem. Soc. (London) **1957**, 381.

[4] BLAU, J. A., W. GERRARD and M. F. LAPPERT: ibid. **1960**, 667.

[5] KÖSTER, R., and K. IWASAKI: Advances in Chemistry **42**, 148 (1964).

Piperidinoboranes, $H_2C \Big\langle \begin{smallmatrix} CH_2 - CH_2 \\ CH_2 - CH_2 \end{smallmatrix} \Big\rangle N - B \Big\langle \begin{smallmatrix} R \\ R \end{smallmatrix}$, and similar substances

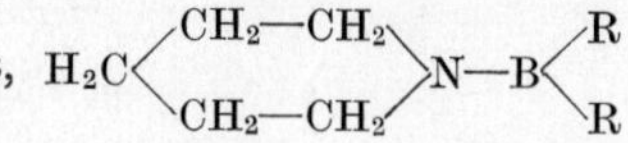

with the nitrogen incorporated in a ring structure have been cited above. A more exotic material is obtained on the reaction of hepta-sulfur imide, S_7NH, with trihalogenoboranes (BCl_3, BBr_3), namely $S_7N - BX_2$ [1].

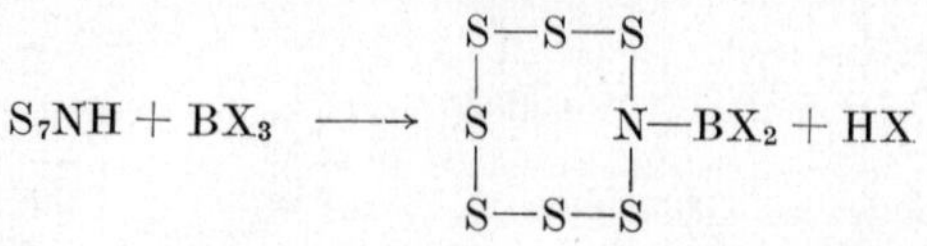

$$
S_7NH + BX_3 \longrightarrow
\begin{matrix}
S - S - S \\
| \qquad\qquad | \\
S \qquad\quad N - BX_2 + HX \\
| \qquad\qquad | \\
S - S - S
\end{matrix}
\qquad (II\text{-}19)
$$

GOUBEAU and GRÄBNER have been able to synthesize (isocyanato)-dimethylborane, $OCN - B(CH_3)_2$ [2]. Similar materials such as the thio-derivative IX were described by LAPPERT and PYSZORA [3]. These thio-derivatives can be reacted with tris(dimethylamino)-borane in the following manner [4].

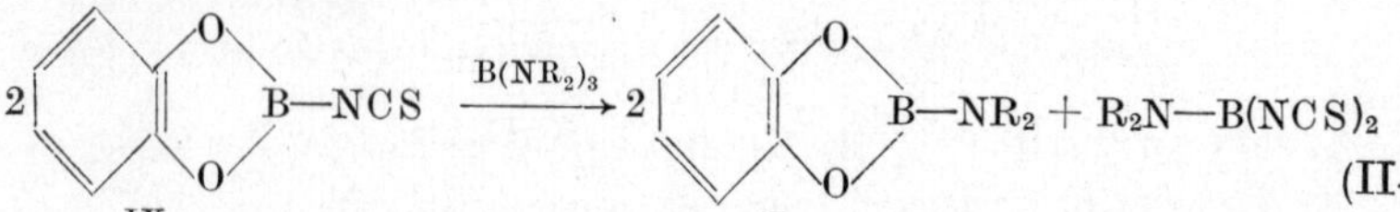

$$ (II\text{-}20) $$

The resultant (amino)diisothiocyanatoborane system is unique, but deri-vatives of type X had been reported previously [5].

N-Silylated aminoboranes have been cited above. An interesting characteristic of these compounds resides in the fact that the N—Si linkage is attacked by halogenoboranes [6]. This reaction was used for the synthesis of diborylamines (see Chapter II-F).

An interesting reaction occurs when isocyanates are added to amino-boranes: the original B—N bond is ruptured and ureidoboranes are formed [7,8].

$$ \rangle B - NR_2 + C_6H_5NCO \rightarrow \rangle B - NC_6H_5 - CO - NR_2 \qquad (II\text{-}21) $$

C. Trimeric Aminoboranes

One of the most characteristic reactions of the borazine ring system, $(-BR - NR'-)_3$, is the facile addition of compounds of the type

---

[1] HEAL, H. G.: J. chem. Soc. (London) **1962**, 4442.

[2] GOUBEAU, J., and H. GRÄBNER: Chem. Ber. **93**, 1379 (1960).

[3] LAPPERT, M. F., and H. PYSZORA: Proc. chem. Soc. (London) **1960**, 350.

[4] HEYING, T. L., and H. D. SMITH JR.: Advances in Chemistry **42**, 201 (1964).

[5] GERRARD, W., M. F. LAPPERT and B. A. MOUNTFIELD: J. chem. Soc. (London) **1959**, 1529.

[6] JENNE, H., and K. NIEDENZU: Inorg. Chem. **3**, 68 (1964).

[7] BEYER, H., J. W. DAWSON, H. JENNE and K. NIEDENZU: J. chem. Soc. (London) **1964**, 2115.

[8] CRAGG, R. H., M. F. LAPPERT and B. P. TILLEY: ibid. **1964**, 2108.

HX (X = halogen, OH, OR) to give 1:3 adducts. WIBERG[1] reported
a number of such adducts of the parent borazine and it was assumed that
they are structurally similar to cyclohexane (XI). However, proof of
this assumption was never presented and the adducts were not even

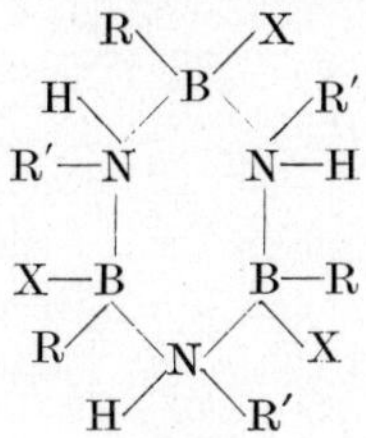

XI

characterized. Actually, the first well-defined trimeric aminoborane[2] was
obtained in 1955 by BISSOT and PARRY[3] in small yield on the pyrolysis
of O,N-dimethylhydroxylamine-borane. The resultant compound, trimeric
methylaminoborane, $[CH_3HNBH_2]_3$, can be prepared more conveniently
and in yields of 80—90% by heating methylamine-borane to 100°.

$$3CH_3H_2N \cdot BH_3 \rightarrow 3H_2 + (CH_3HNBH_2)_3 \qquad (II\text{-}22)$$

The identity of the compound was proved by molecular weight determination and chemical analysis, and by its decomposition at 200° into
hydrogen and N-trimethylborazine, $(—BH—NCH_3—)_3$. The material
does not display the usual instability of monomeric aminoboranes
towards hydrolysis.

In 1957 BURG[4] described a compound to which he gave the formula
$[(CH_3)_2N]_3B_3H_4$ and which he obtained through the reaction of penta-
borane(9), B_5H_9, with an excess of dimethylaminoborane, $(CH_3)_2N—BH_2$.
Two years later, however, a three-dimensional X-ray diffraction study[5]
indicated that this compound is really the trimeric dimethylaminoborane, $[(CH_3)_2N—BH_2]_3$. The compound consists of a cyclic system
of type XI of alternating BH_2 and $N(CH_3)_2$ groups with a B—N bond
distance of 1.59 $\pm$ 0.02 Å[6]. In an independent [11]B nuclear magnetic
resonance study of the material[7], a 1:2:1 triplet was observed suggesting
three equivalent BH_2 groups. It was also demonstrated that the compound
is formed on heating the corresponding dimer with pentaborane(9) or
other higher boron hydrides.

[1] WIBERG, E.: Naturwissenschaften **35**, 182 212 (1948).
[2] SCHAEFFER, R.[8] has proposed the name cycloborazane for this system.
Substitutions would be indicated as they are for borazines, nitrogen atoms being
in the 1, 3, 5 positions. In the following the designation "trimeric aminoboranes"
has been preferred.
[3] BISSOT, T. C., and R. W. PARRY: J. Amer. chem. Soc. **77**, 3481 (1955).
[4] BURG, A. B.: ibid. **79**, 2129 (1957).
[5] TREFONAS, L. M., and W. N. LIPSCOMB: ibid. **81**, 4435 (1959).
[6] TREFONAS, L. M., F. S. MATHEWS and W. N. LIPSCOMB: Acta crystallogr.
[Copenhagen] **14**, 273 (1961).
[7] CAMPBELL, G. W., and L. JOHNSON: J. Amer. chem. Soc. **81**, 3800 (1959).
[8] GAINES, D. F., and R. SCHAEFFER: ibid. **85**, 395 (1963).

More recent investigations of trimeric aminoboranes by R. SCHAEFFER and his coworkers follow the original work of WIBERG who had demonstrated that, at low temperatures, borazine absorbs hydrogen chloride to form a 1:3 adduct $B_3N_3H_6 \cdot 3HCl$. This adduct was then treated with sodium

$$B_3N_3H_6 \cdot 3\,HCl$$

hydroborate and the trimeric aminoborane $(-BH_2-NH_2-)_3$ was isolated and identified[1].

$$2\,B_3N_3H_6 \cdot 3\,HCl + 6\,NaBH_4 \rightarrow 2\,B_3N_3H_{12} + 3\,B_2H_6 + 6\,NaCl \quad \text{(II-23)}$$

This compound is reasonably stable toward atmospheric attack, does not react with cold water, is soluble in polar solvents and has no measurable vapor pressure at temperatures up to 150°, indicating the great intermolecular forces resulting from the alternating positive and negative charges within the ring. Independent molecular weight studies substantiated the trimeric structure[2].

Another approach to the preparation of trimeric aminoboranes was reported by SHORE and coworkers[3]. They found that the diammoniate of diborane (see Chapter I-C) reacts with strong bases such as sodium amide or sodium acetylide in liquid ammonia to produce crystalline species of the formula $[BH_2NH_2]_n$. It appears that the base acts to abstract a proton from a coordinated ammonia moeity of the cation $[BH_2(NH_3)_2]^{\oplus}$ to form $H_3N \cdot BH_2NH_2$. The latter loses ammonia in a self-association process to yield $[BH_2NH_2]_n$. It is possible to isolate 15% of the trimer, $n = 3$, from the crystalline product. The larger portion of the product, however, is higher polymeric. A variety of reports exist on the formation of polymeric aminoboranes[4-8]. Such materials have normally been considered rather unattractive. The studies of SHORE, however,

---

[1] DAHL, G. H., and R. SCHAEFFER: J. Amer. chem. Soc. **83**, 3032 (1961).

[2] SHORE, S. G., and C. W. HICKAM: Inorg. Chem. **2**, 638 (1963).

[3] SHORE, S. G., C. W. HICKAM and K. W. BÖDDEKER: Preprints of Papers, International Symposium on Boron-Nitrogen Chemistry, Durham, N.C., USA, April 1963. See also: SHORE, S.G., K.W. BÖDDEKER, C.W. HICKAM Jr. and D.R. LEAVERS: Abstr. of Papers, 148th National Meeting of the American Chemical Society, Chicago, Ill., **1964**, p. 22—0.

[4] SCHLESINGER, H. I., D. M. RITTER and A. B. BURG: J. Amer. chem. Soc. **60**, 2297 (1938).

[5] WIBERG, E., and P. BUCHHEIT: Z. anorg. allg. Chem. **256**, 285 (1948).

[6] SCHAEFFER, G. W., and L. J. BASILE: J. Amer. chem. Soc. **77**, 331 (1955).

[7] SCHAEFFER, G. W., M. D. ADAMS and F. J. KOENIG: ibid. **78**, 725 (1956).

[8] HOUGH, W. V., and G. W. SCHAEFFER: U.S. Patent 2,869,171 (1957).

seem to indicate that, although these materials are of higher molecular weight, they might not necessarily be highly polymeric. Preliminary results indicate a $[BH_2NH_2]_{10}$ to be a major species in the cited reaction.

In view of former considerations, it appears that the preparation of trimeric aminoboranes is best achieved by an addition reaction to borazines. Thus a 48% yield of trimeric methylaminoborane, $[CH_3HN—BH_2]_3$, was obtained through addition of hydrogen chloride to N-trimethylborazine and subsequent hydrogenation[1]. In addition, the same material was isolated as an intermediate in the preparation of N-trimethylborazine by the method of SCHAEFFER and ANDERSON[2].

Trimeric methylaminoborane can be separated into two structural isomers in about a 7:3 ratio[1]. The ^{11}B nuclear magnetic resonance spectra of both isomers are identical; they consist of the expected 1:2:1 triplet. However, there are characteristic differences in the infrared and proton magnetic resonance spectra. The X-ray powder pattern of one of the isomers can be indexed on an orthorhombic cell closely similar to that found for the trimeric dimethylaminoborane. The two isomers of methylaminoborane trimer interconvert slowly in liquid ammonia at room temperature. The calculated equilibrium constant for this interconvention is 2.28, corresponding to a free energy difference of —0.51 kcal./mole, favoring the symmetrical isomer in which all methyl groups are equatorial; the unsymmetrical material has one axial methyl group and both isomers are probably in the chair shape[3], in agreement with some theoretical considerations[4]. The unsymmetrical isomer forms a 2,4-dichloro substitution product with hydrogen chloride at 0°; a 2,4,6-trichloro derivative alone is obtained from the symmetrical compound[5]. Such observations are in agreement with predictions based on analogies drawn from the corresponding cyclohexanes. The acid-catalyzed methanolysis of the unsymmetrical trimeric methylaminoborane produces a 2-methoxy derivative at 0°. Under the same conditions, the symmetrical isomer yields a mixture of mono-, di- and trisubstituted compounds.

The two isomers of trimeric methylaminoborane

[1] GAINES, D. F., and R. SCHAEFFER: J. Amer. chem. Soc. 85, 395 (1963).

[2] SCHAEFFER, G. W., and E. R. ANDERSON: ibid. 71, 2143 (1949).

[3] BUTTLAR, R. O., D. F. GAINES and R. SCHAEFFER: Preprints of Papers, International Symposium on Boron-Nitrogen Chemistry, Durham, N.C., U.S.A. 1963.

[4] HOFFMANN, R.: Advances in Chemistry 42, 78 (1964).

[5] GAINES, D. F., and R. SCHAEFFER: J. Amer. chem. Soc. 85, 3592 (1963).

D. Bisaminoboranes

Prior to 1956 only one linear bisaminoborane was known. On heating dimethylaminoborane, $(CH_3)_2N—BH_2$, with dimethylamine[1] or trimethylamine[2] for ten hours at 200°, one obtains bis(dimethylamino)borane, $[(CH_3)_2N]_2BH$, among other products. The same compound is formed by the reversible disproportionation of dimethylaminoborane[2] or by the reduction of bis(dimethylamino)chloroborane, $[(CH_3)_2N]_2BCl$, with lithium aluminium hydride[3]. Bis(dimethylamino)chloroborane has been described by WIBERG and SCHUSTER[4] and probably was obtained earlier[5]. G. E. COATES[3] and SKINNER and SMITH[6] studied the thermochemistry of this compound, and WIBERG[7] obtained bis(dimethylamino)bromoborane on dimethylaminolysis of tribromoborane. In recent years, due mainly to efforts to synthesize boron-boron linkages via a WURTZ-FITTIG reaction (see Chapter II-H), bisaminohalogenoboranes have been studied in more detail and a variety of synthetic approaches to the bisaminoborane system were discovered. Disproportionation of trisaminoborane and trihalogenoborane[8] has been found to be an effective method for the synthesis of bisaminoboranes. MIKHAILOV[9] has reported the aminolysis of aminomercaptylboranes.

$$2B(NR_2)_3 + BX_3 \rightarrow 3(R_2N)_2BX \tag{II-24}$$
$$(R_2N)BH(SR') + R_2NH \rightarrow (R_2N)_2BH + R'SH \tag{II-25}$$

At 100—150°, a dialkylamine can react with a trimethylamine-alkylborane in the presence of catalytic amounts of ammonium ion to afford bis(dialkylamino)alkylboranes[10]. These derivatives have also been obtained through the interaction of certain other boron compounds, e.g. B-organoboroxines, with tris(dimethylamino)alane, $Al[N(CH_3)_2]_3$ [11]. The aminolysis of organodichloroboranes was studied first by MIKHAILOV and coworkers[12, 13]. Apparently the choice of solvents is very important

[1] WIBERG, E., and A. BOLZ: Z. anorg. allg. Chem. **257**, 131 (1948).
[2] BURG, A. B., and C. L. RANDOLPH: J. Amer. chem. Soc. **73**, 953 (1951).
[3] COATES, G. E.: J. chem. Soc. (London) **1950**, 3481.
[4] WIBERG, E., and K. SCHUSTER: Z. anorg. allg. Chem. **213**, 77, 89 (1933).
[5] MICHAELIS, A., and K. LUXEMBOURG: Ber. dtsch. chem Ges. **29**, 710 (1896).
[6] SKINNER, H. A., and N. B. SMITH: J. chem. Soc. (London) **1954**, 2324.
[7] WIBERG, E., and W. STURM: Z. Naturforsch. 8b, 689 (1953).
[8] BROTHERTON, R. L., A. L. McCLOSKEY, L. L. PETTERSON and H. STEINBERG: J. Amer. chem Soc. **82**, 6242 (1960).
[9] MIKHAILOV, B. M., and W. A. DOROCHOV: J. Gen. Chem. USSR. **31**, 750 (1961).
[10] HAWTHORNE, M. F.: J. Amer. chem. Soc. **83**, 2671 (1961).
[11] RUFF, J. K.: J. org. Chemistry **27**, 1020 (1962).
[12] MIKHAILOV, B. M., and P. M. ARONOVICH: Bull. Acad. Sci. USSR., Div. Chem. Sci. **1957**, 1146.
[13] MIKHAILOV, B. M., A. N. BLOKHINA and T. V. KOSTROMA: J. Gen. Chem. USSR. **29**, 1483 (1959).

if high yields are to be obtained in this reaction [1-3]. (Bisamino)alkoxy-boranes have been prepared by either of the three illustrated methods[4].

$$ROBCl_2 + 4R_2'NH \rightarrow ROB(NR_2')_2 + 2[R_2'NH_2]Cl \quad \text{(II-26)}$$

$$ROH + B(NR_2')_3 \rightarrow ROB(NR_2')_2 + R_2'NH \quad \text{(II-27)}$$

$$ROH + (R_2'N)_2BCl + (C_2H_5)_3N \rightarrow ROB(NR_2')_2 + [(C_2H_5)_3NH]Cl \quad \text{(II-28)}$$

As cited above, the primary interest in bisaminoborane chemistry is centered around the halogen derivatives, $(R_2N)_2BX$, which have been widely used in the synthesis of diborane(4) derivatives (see Chapter II-H) and have since been studied in great detail. The halogen is readily replaced by organic groups when a representative compound is reacted with metal-organics. In like manner, lithium hydride reacts in ether solution as shown in the equation[5]:

$$(R_2N)_2BX + LiH \rightarrow (R_2N)_2BH + LiX \quad \text{(II-29)}$$

It is of interest to compare the reactions of bisaminoboranes, $(R_2N)_2BR'$, with hydrogen halide at room temperature. This reaction has been found to depend primarily upon the nature of R' as was studied mainly by Nöth and his coworkers with bis(dimethylamino)boranes. They were able to demonstrate that two major reaction patterns are followed: If the substituent R' is $N(CH_3)_2$ [6], SiR_3 [7], $B[N(CH_3)_2]_2$ [8] or an alkoxy group[4], cleavage of a boron-nitrogen bond by hydrogen chloride occurs according to the following equation:

$$(R_2N)_2BR' + 2HCl \rightarrow R_2N-B\underset{Cl}{\overset{R'}{\diagdown}} + [R_2NH_2]Cl \quad \text{(II-30)}$$

With $R' = H$ or Cl, however, a different reaction path is followed, whereby a borazylammonium salt is formed (see also Chapter I-C)[5, 6]:

$$(R_2N)_2BR' + 2HCl \longrightarrow \begin{bmatrix} R_2HN\diagdown & \diagup R' \\ & B \\ R_2HN\diagup & \diagdown Cl \end{bmatrix}^{\oplus} Cl^{\ominus} \quad \text{(II-31)}$$
$$\text{XII}$$

These salts are stable at room temperature and, in agreement with an earlier observation by Burg and Slota[9], an excess of hydrogen chloride

[1] Niedenzu, K., H. Beyer and J. W. Dawson: Inorg. Chem. 1, 738 (1962).
[2] Nöth, H., and P. Fritz: Z. anorg. allg. Chem. 322, 297 (1963).
[3] Burch, J. E., W. Gerrard and E. F. Mooney: J. chem. Soc. (London) 1962, 2200.
[4] Gerrard, W., M. F. Lappert and C. A. Pearce: ibid. 1957, 381.
[5] Nöth, H., W. A. Dorochov, P. Fritz and F. Pfab: Z. anorg. allg. Chem. 318, 293 (1962).
[6] Nöth, H., and S. Lukas: Chem. Ber. 95, 1505 (1962).
[7] Nöth, H., and G. Hollerer: Angew. Chem. 74, 718 (1962).
[8] Nöth, H., and W. Meister: Z. Naturforsch. 17b, 714 (1962).
[9] Burg, A. B., and J. P. Slota: J. Amer. chem. Soc. 82, 2148 (1960).

does not attack the B—H bond. If R' is an alkyl group, other than methyl, interaction with hydrogen chloride follows the course of eq. II-30. In contrast to this occurrence, the reaction of bis(dimethylamino)methyl-borane with HCl yields a salt in accordance with eq. II-31[1]. The resultant salt is unstable to heat at 170—180° (eq. II-32, R=X=CH$_3$).

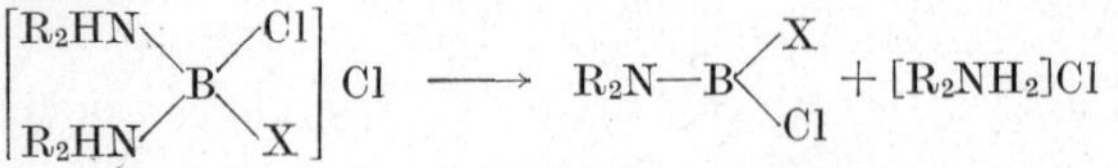

$$\left[\begin{array}{c} R_2HN \\ \\ R_2HN \end{array} B \begin{array}{c} Cl \\ \\ X \end{array}\right] Cl \longrightarrow R_2N-B \begin{array}{c} X \\ \\ Cl \end{array} + [R_2NH_2]Cl \qquad (II\text{-}32)$$

In addition to the above observations, the nature of the substituent R at the nitrogen seems to play an important part in the course of the reaction. If R is an alkyl group other than methyl, the action of hydrogen chloride upon bisaminoboranes will follow the pattern illustrated in eq. II-30.

If X=H, (eq. II-32) additional hydrogen chloride can add across the remaining boron-nitrogen bond of the resultant monoaminoborane to produce dialkylamine-dichloroboranes as final products (eq. II-33).

$$(R_2N)_2BH + 3\,HCl \rightarrow R_2HN \cdot BHCl_2 + [R_2NH_2]Cl \qquad (II\text{-}33)$$

The variation in the above reaction patterns upon replacement of methyl by ethyl groups indicates that the size of substituents may play an important part in the thermal, and most likely also in the hydrolytic stability of borazylammonium salts. Therefore, it is surprising that hydrogen bromide reacts analogously to hydrogen chloride in accordance with the pattern outlined above. This behavior seems to indicate that spatial requirements are a *contributing*, but not a *major* factor, in improving the stability of borazylammonium salts.

The interaction of bis(dimethylamino)methylborane, $[(CH_3)_2N]_2BCH_3$, with hydrogen iodide appears to be most interesting. Whereas bis(dimethylamino)butylborane reacts with hydrogen iodide according to the pattern indicated by eq. II-30, bis(dimethylamino)methylborane consumes only one mole of hydrogen iodide and, in toluene solution, quantitatively precipitates a 1:1 adduct. This adduct exhibits a medium to strong infrared absorption at 1576 cm.$^{-1}$, a frequency assigned to an unsymmetrical B—N stretching mode. Since this occurrence would indicate extremely strong B—N double bond character, the following structure was formulated[1]:

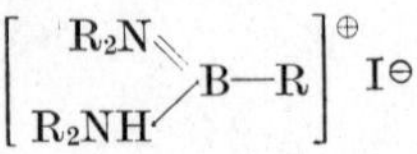

$$\left[\begin{array}{c} R_2N \\ \\ R_2NH \end{array} B-R\right]^{\oplus} I^{\ominus}$$

XIII

The structure of this resonance-stabilized cation XIII finds additional support in the observation that the addition of hydrogen chloride to XIII results in the disappearance of the assigned B—N absorption and the

---

[1] Nöth, H., and P. Fritz: Z. anorg. allg. Chem. **322,** 297 (1963).

formation of a borazylammonium salt. Further support for the formulation of the cation XIII may be seen in a similar reaction of tris(dimethylamino)borane with hydrogen iodide (eq. II-33).

$$R_2N\!\!\diagdown\!\!\!\!\!\diagup^{\displaystyle B\!\!-\!\!NR_2} + 2\,HI \quad\longrightarrow\quad \left[\begin{array}{c}R_2HN\!\!\diagdown\\R_2HN\!\!\diagup\end{array}\!\!\!\!\!B\!\!=\!\!NR_2\right]^{\oplus\oplus} I_2^{\ominus\ominus} \qquad (II\text{-}33)$$
$$\text{VIX}$$

The resultant salt, XIV, shows a B—N absorption in the infrared similar to that of XIII, which appears at 1558 cm.$^{-1}$.

Bisaminoboranes are known only as monomers. Consequently, the chemical shift of the ^{11}B nuclear magnetic resonance has been found at negative values of about $-20(\delta\times10^{-6})$ [1,2].

Bisaminoboranes have recently been used in the preparation of a variety of boron-nitrogen-carbon heterocyles which will be described later (see Chapter V-B). Some typical examples of variously substituted bisaminoboranes are compiled in Table II-4.

Table II-4. *Bisaminoboranes*

	b.p. (mm.) °C	References
$[(CH_3)_2N]_2B\!\!-\!\!CH_3$	29—32 (15)	7
$[(CH_3)_2N]_2B\!\!-\!\!C_2H_5$	33—35 (10)	7
$[(CH_3)_2N]_2B\!\!-\!\!n\text{-}C_4H_9$	50—51 (5)	2, 7
$[(CH_3)_2N]_2B\!\!-\!\!t\text{-}C_4H_9$	66 (20)	2
$[(C_2H_5)_2N]_2B\!\!-\!\!n\text{-}C_4H_9$	77 (0.3)	4
$[(CH_3)HN]_2B\!\!-\!\!C_6H_5$	106—107 (16)	3
$[(CH_3)_2N]_2B\!\!-\!\!C_6H_5$	59 (3)	2, 5, 6
$[(C_2H_5)HN]_2B\!\!-\!\!C_6H_5$	106—108 (10)	3
$[(C_2H_5)_2N]_2B\!\!-\!\!C_6H_5$	70 (2)	5, 6
$[(n\text{-}C_4H_9)HN]_2B\!\!-\!\!C_6H_5$	118—119 (0.5)	3

A most interesting reaction of bis(dimethylamino)chloroborane occurs when it is treated with the alkali salts of transition metal carbonyls[8].

$$[(CH_3)_2N]_2BCl + NaMn(CO)_5 \rightarrow [(CH_3)_2N]_2B\!\!-\!\!Mn(CO)_5 + NaCl \quad (II\text{-}35)$$
$$\text{XV}$$

Compounds of type XV with a stable σ-bond between boron and a metal are new. A variety of chloroboranes can participate in this reaction;

---

[1] PHILLIPS, W. D., H. C. MILLER and E. L. MUETTERTIES: J. Amer. chem. Soc. **81**, 4496 (1959).

[2] RUFF, J. K.: J. org. Chemistry **27**, 1020 (1962).

[3] BURCH, J. E., W. GERRARD and E. F. MOONEY: J. chem. Soc. (London) **1962**, 2200.

[4] HAWTHORNE, M. F.: J. Amer. chem. Soc. **83**, 2671 (1961).

[5] MIKHAILOV, B. M., and N. S. FEDOTOV: Bull. Acad. Sci. USSR., Div. Chem. Sci. **1956**, 1511.

[6] NIEDENZU, K., H. BEYER and J. W. DAWSON: Inorg. Chem. **1**, 738 (1962).

[7] NÖTH, H., and P. FRITZ: Z. anorg. allg. Chem. **322**, 297 (1963).

[8] NÖTH, H., and G. SCHMID: Angew. Chem. **75**, 861 (1963).

so far, the following products have been characterized:

$[(CH_3)_2N]_2B$—$Mn(CO)_5$	decomp. 60°
$(CH_3)_2NB$—$[Mn(CO)_5]_2$	decomp. 85°
$(C_6H_5)[(CH_3)_2N]B$—$Mn(CO)_5$	m.p. 146—150°
$[(C_6H_5)_2BMn]_2$	m.p. 216—218°
$[(CH_3)_2N]_2B$—$Mn(CO)_4P(C_6H_5)_3$	m.p. 125°
$(C_4H_9)_2B$—$Mn(CO)_4P(C_6H_5)_3$	m.p. 100° (decomp.)
$(C_6H_5)_2B$—$Mn(CO)_4P(C_6H_5)_3$	m.p. 120°

Also, the interaction of bisaminoboranes with isocyanates and isothiocyanates[1, 2] appears to be of interest. With isocyanates, bisureido-boranes, XVI, are obtained through cleavage of the (original) boron-nitrogen bonds:

$$C_6H_5B(NR_2)_2 + 2C_6H_5NCO \rightarrow C_6H_5B\Big\langle \begin{array}{l} N(C_6H_5)\text{—}CO\text{—}NR_2 \\ N(C_6H_5)\text{—}CO\text{—}NR_2 \end{array} \qquad (II\text{-}36)$$
$$XVI$$

Isothiocyanates react with bisaminoboranes in similar fashion to yield bis(thioureido)boranes. However, such compounds appear to be stable only if the terminal nitrogens are doubly substituted. If this is not the case, reaction according eq. II-37 seems to be preferred.[2]

$$C_6H_5B(NHR)_2 + 2C_6H_5NCS \rightarrow C_6H_5B(NHC_6H_5)_2 + 2RNCS \qquad (II\text{-}37)$$

a. Bis(dimethylamino)chloroborane by Disproportionation[3]

A solution of 143 g. (1 mole) of tris(dimethylamino)borane in 250 cc. of dry hexane is cooled in an ice bath. A cold solution of 41 ml. (0.5 mole) of trichloro-borane in 200 cc. of dry hexane is added with stirring over a period of 20—30 minutes. The solvent is evaporated and the residue distilled in vacuum. Yield: about 90%, b.p. 38° (10 mm.).

b. Bis(dimethylamino)borane[4]

A quantity, 4.5 g. (640 mmole) of powdered lithium hydride is added slowly to a solution of 62.5 g. (464 mmole) of bis(dimethylamino)chloroborane in 100 cc. of dry ether. Upon completion of the exothermic reaction, the mixture is refluxed for two hours and filtered in an inert atmosphere; the ether is removed through an effective column. (A 25 cm. silver mantle column filled with stainless steel helices is recommended) Fractionation of the residue yields 36.6 g. (79%) of bis(dimethylamino)borane, b.p. 106—108° (720 mm.).

Note[3]: The powdered lithium hydride is prepared by grinding 100 g. of lithium hydride in 1000 cc. of dry ether for eight hours in a ball mill. A suspension of the highly active hydride in the ether is obtained and the desired quantity of lithium hydride is added in the form of an ethereal suspension to the reaction vessel through a dropping funnel.

c. Bis(dimethylamino)phenylborane[5]

A solution of 100 g. (2.22 mole) of anhydrous dimethylamine in 1200 cc. of dry hexane is cooled in a Dry Ice-methanol bath. A mixture of 79.4 g. (0.5 mole) of phenyldichloroborane and 300 cc. of dry hexane is added with stirring over a period of thirty minutes. The cooling bath is removed, and the mixture is

[1] BEYER, H., J. W. DAWSON, H. JENNE and K. NIEDENZU: J. chem. Soc. (London) 1964, 2115.

[2] CRAGG, R. H., M. F. LAPPERT and B. P. TILLEY: ibid. 1964, 2108.

[3] FRITZ, P.: personal communication.

[4] NÖTH, H., W. A. DOROCHOV, P. FRITZ and F. PFAB: Z. anorg. allg. Chem. 318, 293 (1962).

[5] NIEDENZU, K., H. BEYER and J. W. DAWSON: Inorg. Chem. 1, 738 (1962).

stirred until it reaches room temperature. It is then filtered and the solvent stripped from the filtrate. Distillation of the residue under vacuum yields 77 g. (87.5%) of bis(dimethylamino)phenylborane, b.p. 59° (3 mm.).

E. Trisaminoboranes

Trisaminoboranes, $B(NRR')_3$, are generally prepared by the addition of trichloroborane to an excess of amine in an inert solvent at low temperatures[1-3].

$$BCl_3 + 6\,HNRR' \rightarrow B(NRR')_3 + 3\,[H_2NRR']Cl \qquad (II\text{-}38)$$

The first member of this series of compounds, tris(dimethylamino)borane, $B[N(CH_3)_2]_3$, was described by WIBERG and SCHUSTER[4] and was obtained by reaction (II-38). Although the reaction was originally effected in the vapor phase, the use of inert solvents as reaction media is most advantageous[5, 6]. Tris(dimethylamino)borane was also obtained as a minor product in the reaction of bis(dimethylamino)chloroborane, $ClB[N(CH_3)_2]_2$, with dimethylzinc[5]. It can be assumed that the trisaminoborane is formed in this reaction by disproportionation of the initial bis(dimethylamino)methylborane in the presence of an excess of the organometallic[7]. Later, tris(dimethylamino)borane was found as a product from the pyrolysis of either bis(dimethylamino)borane[8], $[(CH_3)_2N]_2BH$, or (dimethylamino)dichloroborane[9], $(CH_3)_2N\text{—}BCl_2$, in the presence of trimethylamine, and also on reacting tribromoborane with an excess of dimethylamine[10]. Tris(phenylamino)borane $B(NHC_6H_5)_3$ was synthesized through the reaction of trichloroborane with aniline[11]. The p-tolyl[12] and p-anisyl[13] homologs were obtained in like manner.

The reaction of trifluoroborane with primary and secondary amines in the presence of either a GRIGNARD reagent[14] or metallic lithium[15] has also been described. The following mechanism was formulated.

$$R_2NH + R'MgX \rightarrow R_2N\text{—}MgX + R'H \qquad (II\text{-}39)$$
$$3\,R_2N\text{—}MgX + BF_3 \rightarrow B(NR_2)_3 + 3\,MgFX \qquad (II\text{-}40)$$

An additional method involves the use of transamination

$$B(NR_2)_3 + 3\,R_2'NH \rightleftharpoons B(NR_2')_3 + 3\,R_2NH \qquad (II\text{-}41)$$

[1] AUBREY, D. W., M. F. LAPPERT and M. K. MAJUMDAR: J. chem. Soc. (London) **1962**, 4088.

[2] AUBREY, D. W., W. GERRARD and E. F. MOONEY: ibid. **1962**, 1786.

[3] GERRARD, W., M. F. LAPPERT and C. A. PEARCE: ibid. **1957**, 381.

[4] WIBERG, E., and K. SCHUSTER: Z. anorg. allg. Chem. **213**, 77, 89 (1933).

[5] COATES, G. E.: J. chem. Soc. (London) **1950**, 3481.

[6] SKINNER, H. A., and N. B. SMITH: ibid. **1954**, 2324.

[7] NÖTH, H., and P. FRITZ: Z. anorg. allg. Chem. **322**, 297 (1963).

[8] BURG, A. B., and C. L. RANDOLPH: J. Amer. chem. Soc. **73**, 953 (1951).

[9] BROWN, C. A., and R. C. OSTHOFF: ibid. **74**, 2340 (1952).

[10] WIBERG, E., and W. STURM: Z. Naturforsch. **8b**, 689 (1953).

[11] JONES, R. G., and C. R. KINNEY: J. Amer. chem. Soc. **61**, 1378 (1939).

[12] KINNEY, C. R., and M. J. KOLBEZEN: ibid. **64**, 1584 (1942).

[13] KINNEY, C. R., and C. L. MAHONEY: J. org. Chemistry **8**, 526 (1943).

[14] DORNOW, A., and H. H. GEHRT: Z. anorg. allg. Chem. **294**, 81 (1958).

[15] KRAUS, C. A., and E. H. BROWN: J. Amer. chem. Soc. **52**, 4414 (1930).

which has also been used for the preparation of unsymmetrical tris-aminoboranes[1], $(R_2N)_2B$—NRR'. Heretofore, such compounds were virtually unknown; only a few derivatives were obtained under sterically favored conditions. For example, a (bisamino)chloroborane, $ClB(NR_2)_2$, in which R is a bulky organic group, will not react with a third identical amine molecule, whereas a sterically less demanding amine is able to afford the unsymmetrical product[2, 3].

In a few selected cases, it has been found possible to convert tri-alkoxyboranes to more complex trisaminoboranes; such structures will be discussed in a later chapter. A patent also claims the formation of tris(phenylamino)borane through the reaction of isocyanates with boric acid[4] (eq. II-42).

$$3\,RNCO + H_3BO_3 \rightarrow B(NHR)_3 + 3\,CO_2 \qquad (II\text{-}42)$$

The physicochemical investigation of trisaminoboranes is not as advanced as the preparative studies. An investigation of the valence force constant of the boron-nitrogen bond in tris(dimethylamino)borane[5] yielded a value of about 5.5×10^{15} dyne/cm. This value is lower than those reported for borazines (6.3) and more than that of monoamino-boranes (ca. 3.5×10^{-5} dyne/cm.). The infrared spectra of various tris-aminoboranes have been recorded[6] and in many cases it is obvious that a considerable degree of double bond character can be ascribed to the boron-nitrogen linkage. The bond order, however, decreases in the series

$$B(NHAryl)_3 > B(NHAlkyl)_3 > B(NAlkyl_2)_3 > B(NAlkylAryl)_3$$

The heat of formation of tris(dimethylamino)borane was determined by SKINNER and SMITH[7] to be -77.1 kcal./mole. From this one can calculate the energy of dissociation of the B—N bond to be 89.7 kcal./mole This value is of the same order as that of the boron-nitrogen bond in borazines.

There is strong evidence that hydrochlorides of the type $B(NR_2)_3 \cdot nHCl$ generally are not stable[8]. Upon reacting an excess of hydrogen chloride with tris(dimethylamino)borane, four moles of the hydrogen halide are absorbed and a borazylammonium salt is formed.

$$B(NR_2)_3 + 4\,HCl \longrightarrow \begin{bmatrix} R_2HN & Cl \\ & B \\ R_2HN & Cl \end{bmatrix}^{\oplus} Cl^{\ominus} + [R_2NH_2]Cl \qquad (II\text{-}43)$$

The structure of this boronium salt conforms to that of XII described above $(R'=Cl)$ and, it is of interest to note that a product of this

[1] BROTHERTON, R. J., and T. BUCKMAN: Inorg. Chem. **2**, 424 (1963).

[2] AUBREY, D. W., M. F. LAPPERT and M. K. MAJUMDAR: J. chem. Soc. (London) **1962**, 4088.

[3] AUBREY, D. W., W. GERRARD and E. F. MOONEY: ibid. **1962**, 1786,

[4] ARIES, R. S.: U.S. Patent 2,931,831 (1960).

[5] BECHER, H. J.: Z. anorg. allg. Chem. **287**, 285 (1956).

[6] AUBREY, D. W., M. F. LAPPERT and H. PYSZORA: J. chem. Soc. (London) **1960**, 5239.

[7] SKINNER, H. A., and N. B. SMITH: ibid. **1953**, 4025.

[8] NÖTH, H., and S. LUKAS: Chem. Ber. **95**, 1505 (1962).

analytical composition was described by WIBERG and SCHUSTER more than thirty years earlier[1]. Tris(diethylamino)borane, however, reacts with an excess of hydrogen chloride to effect cleavage of all normal covalent boron-nitrogen bonds and to yield an amine-borane.

$$B(NR_2)_3 + 5\,HCl \rightarrow Cl_3B \cdot NHR_2 + 2[R_2NH_2]Cl \qquad (II\text{-}44)$$

This difference in the behavior of tris(dialkylamino)boranes towards attack by hydrogen chloride may be the result of a steric effect. Indeed, as shown by LAPPERT and coworkers[2], steric effects seem to be of major importance in trisaminoborane chemistry. They noted that sterically hindered bisaminochloroboranes exhibit a high order of stability with respect to attack by additional hindered amines. Furthermore, hindered unsymmetrical trisaminoboranes are, in general, surprisingly stable towards disproportionation; specifically, tris(methylphenylamino)-borane was found to be extremely resistant to nucleophilic (but not to electrophilic) attack.

Tris(diethylamino)borane[3]

A solution of 4.85 g. of trichloroborane in 25 cc. of dry pentane is cooled to −40° and added dropwise to a solution of 18.6 g. of diethylamine in 100 cc. of the same solvent at − 80°. After filtration and evaporating the filtrate free of solvent, the residue is distilled in vacuum to yield 3.59 g. (38%) of tris(diethylamino)borane, b.p. 50−55° (0.4 mm.).

F. Diborylamines

Diborylamines, XVII, in which two boron atoms are bonded to a central nitrogen atom are exceedingly rare. They were first reported by NÖTH[4], who obtained these compounds through reacting dialkylchloroboranes with bis(trimethylsilyl)amine.

$$2\,R_2BCl + [(CH_3)_3Si]_2NH \rightarrow R_2B\text{---}NH\text{---}BR_2 + 2(CH_3)_3SiCl \qquad (II\text{-}45)$$
$$XVII$$

KÖSTER[5] reported analogous compounds with a B—N—B grouping in which the boron is incorporated into an aliphatic ring. These were formed from the thermal decomposition of a mixture of bisborolanes (XVIII) and aminoborolanes (eq. II-46).

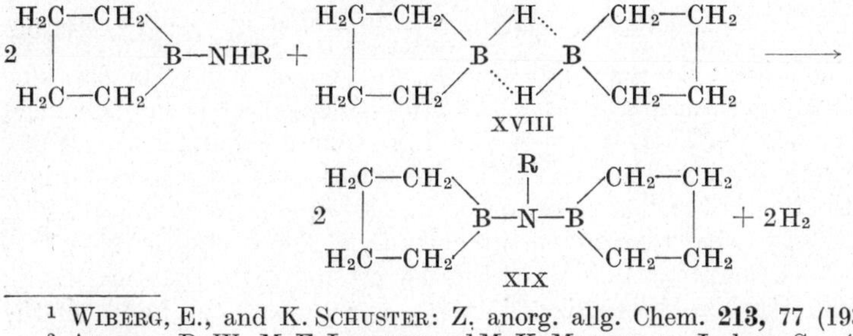

[1] WIBERG, E., and K. SCHUSTER: Z. anorg. allg. Chem. **213**, 77 (1933).
[2] AUBREY, D. W., M. F. LAPPERT and M. K. MAJUMDAR: J. chem. Soc. (London) **1962**, 4088.
[3] GERRARD, W., M. F. LAPPERT and C. A. PEARCE: ibid. **1957**, 381.
[4] NÖTH, H.: Z. Naturforsch. **16 b**, 618 (1961).
[5] KÖSTER, R., and K. IWASAKI: Advances in Chemistry **42**, 148 (1964).

LAPPERT[1] reported the preparation of XX by aminolysis of bis(t-butyl-amino)chloroborane, $[RHN]_2BCl$, with tertiary butylamine.

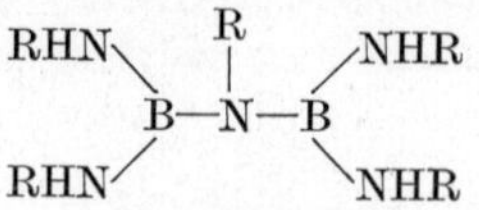

$$
\begin{array}{c}
RHN \\ \quad \\ RHN
\end{array}
\underset{B-N-B}{\overset{R}{\diagup\;\;\diagdown}}
\begin{array}{c}
NHR \\ \quad \\ NHR
\end{array}
\quad R=t\text{-butyl}
$$

XX

A more detailed investigation of this class of compounds has been reported by JENNE and NIEDENZU[2]. In the first step of reaction II-45, a N-silylated aminoborane is formed.

$$R_2BCl + [(CH_3)_3Si]_2NH \rightarrow R_2B-NH-Si(CH_3)_3 + (CH_3)_3SiCl \quad (II\text{-}47)$$

With an excess of the diorganochloroborane, the diborylamine XVII is obtained along with the elimination of additional trimethylchloro-silane (eq. II-48).

$$R_2B-NH-Si(CH_3)_3 + ClBR_2 \rightarrow R_2B-NH-BR_2 + Cl-Si(CH_3)_3 \quad (II\text{-}48)$$

Similarly, on reacting aminochloroboranes with the bis(trimethylsilyl)-amine, a monosilylated bisaminoborane, XXI, appears as the primary

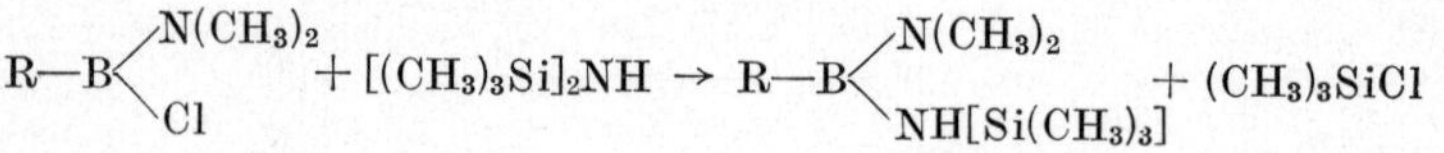

$$
R-B\underset{Cl}{\overset{N(CH_3)_2}{\diagup}} + [(CH_3)_3Si]_2NH \rightarrow R-B\underset{NH[Si(CH_3)_3]}{\overset{N(CH_3)_2}{\diagup}} + (CH_3)_3SiCl \quad (II\text{-}49)
$$

XXI

product. This in turn reacts with an excess of the aminochloroborane to yield the diborylamine XXII. Finally, a material as formulated in

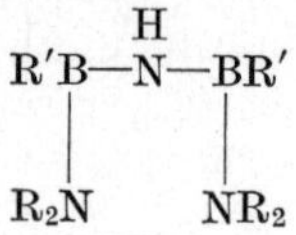

$$
\begin{array}{c}
\overset{H}{R'B-N-BR'} \\ \quad \\ R_2N \quad\quad NR_2
\end{array}
\qquad
\left[
\begin{array}{c}
CH_2-O \\ H_2C \qquad\qquad B-NH \\ CH_2-O
\end{array}
\right]_2
$$

XXII 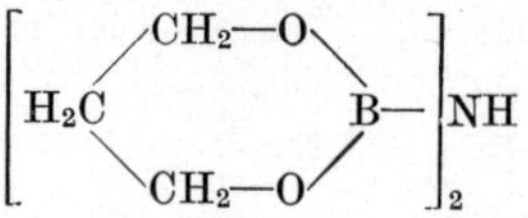   XXIII

XXIII has been cited in the literature[3] but no characterizing data are available.

Diborylamines of type XVII are relatively unstable. They decompose easily according to:

$$3(R_2B)_2NH \rightarrow 3BR_3 + (-BR-NH-)_3 \quad (II\text{-}50)$$

to yield borazines; this decomposition is catalyzed by acids. The borolane groups in diborylamines of type XIX, (p. 73), have not been exchanged to yield acyclic derivatives of type XVII. For example, on treatment of a diborylamine having structure XIX with tetraalkyldiborane, one might expect to effect an exchange of the borane groups and to produce XVII. However, extensive rearrangement occurs yielding cyclic B—N—C de-rivatives (see Chapter V) along with other products. On the other hand, transboronation can be realized in those cases where the boron of the

[1] LAPPERT, M. F., and M. K. MAJUMDAR: Proc. Chem. Soc. (London) 1963, 88.
[2] JENNE, H., and K. NIEDENZU: Inorg. Chem. 3, 68 (1964).
[3] FINCH, A., P. J. GARDNER, J. C. LOCKHART and E. J. PEARN: J. chem. Soc. (London) 1962, 1428.

resultant diborylamine is incorporated in a carbon ring as in **XIX**. Other exchange reactions of boron-attached groups of diborylamines will be described later (see Chapter IV-B).

Only eleven diborylamines have so for been characterized. They are listed in Table II-5.

Table II-5. *Diborylamines*

	b.p. (mm.) °C	References
$[(n\text{-}C_3H_7)_2B]_2NH$	80—84 (3)	7
$[(n\text{-}C_4H_9)_2B]_2NH$	75—78 (0.01)	7
$[(n\text{-}C_5H_{11})_2B]_2NH$	86—87 (2)	6
$\left[\begin{array}{c} C_6H_5 \\ (CH_3)_2N \end{array}\!\!>\!B-\right]_2 NH$	158—160 (3)	2
$\left[\begin{array}{c} Cl \\ (t\text{-}C_4H_9)HN \end{array}\!\!>\!B-\right]_2 N\!-\!t\text{-}C_4H_9$	82—86 (0.02)	4
$\left[\begin{array}{c} (t\text{-}C_4H_9)HN \\ (t\text{-}C_4H_9)HN \end{array}\!\!>\!B-\right]_2 N\!-\!t\text{-}C_4H_9$	98—100 (0.01)	4
$\left[\begin{array}{c} CHCH_3\!-\!CH_2 \\ | \\ CH_2\!-\!-\!CH_2 \end{array}\!\!>\!B-\right]_2 NC_2H_5$	121—123 (11.5)	3
$\left[\begin{array}{c} CHCH_3\!-\!CH_2 \\ | \\ CH_2\!-\!-\!CH_2 \end{array}\!\!>\!B-\right]_2 NC_6H_5$	98.5—100 (0.3)	3
$\left[\begin{array}{c} CHCH_3\!-\!CH_2 \\ | \\ CH_2\!-\!-\!CH_2 \end{array}\!\!>\!B-\right]_2 N\!-\!CH_2C_6H_5$	137—138 (0.4)	3
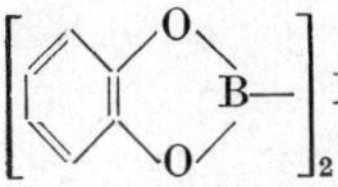 $\left[\begin{array}{c} O \\ O \end{array}\!\!>\!B-\right]_2 N\!-\!CO\!-\!NH\!-\!t\text{-}C_4H_9$	m.p. 198°	1
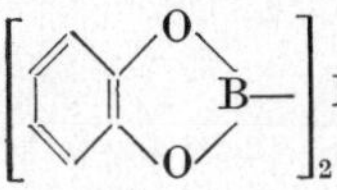 $\left[\begin{array}{c} O \\ O \end{array}\!\!>\!B-\right]_2 NH$	subl. 200° (2 mm.)	5

Recently[5] the first example of a triborylamine was reported. Reacting di(1,3,2-dioxoborolidyl)amine with 1-chloro-1,3,2-dioxoborolidine, in the

[1] Cragg, R. H., M. F. Lappert and B. P. Tilley: J. chem. Soc. (London) **1964**, 2108.

[2] Jenne, H., and K. Niedenzu: Inorg. Chem. **3**, 68 (1964).

[3] Köster, R., and K. Iwasaki: Advances in Chemistry **42**, 148 (1964).

[4] Lappert, M. F., and M. K. Majumdar: Proc. chem. Soc. (London) **1963**, 88.

[5] Lappert, M. F., and G. Srivastava: Proc. chem. Soc. (London) **1964**, 120.

[6] Niedenzu, K., and coworkers: unpublished results.

[7] Nöth, H.: Z. Naturforsch. **16 b**, 618 (1961).

presence of triethylamine, the triborylamine was obtained as a crystalline solid, insoluble in common organic solvents, and readily hydrolized.

$$\left[\begin{array}{c} \overset{O}{\underset{O}{\bigvee}} B- \end{array}\right]_2 NH + Cl-B\overset{O}{\underset{O}{\bigvee}} \xrightarrow{N(C_2H_5)_3}$$

$$\left[\begin{array}{c} \overset{O}{\underset{O}{\bigvee}} B- \end{array}\right]_3 N + [(C_2H_5)_3NH]Cl \qquad (II\text{-}51)$$

G. Hydrazino- and Azido-Boranes

It has been only during the last few years that hydrazinoboranes have attracted attention. This is surprising, since they exhibit some interesting aspects from a theoretical point of view. Boron-nitrogen double bonding, as established for the aminoborane system, might result in conjugation effects; symmetrical N,N'-diborylhydrazines (XXIV) might well offer interesting comparisons to the butadiene system[1].

$$\underset{(a)}{\overset{\diagdown}{\diagup}B-\overset{\cdot\cdot}{N}-\overset{\cdot\cdot}{N}-B\overset{\diagup}{\diagdown}} \qquad \underset{(b)}{\overset{\diagdown}{\diagup}B\overset{\cdot}{=}N-N\overset{\cdot}{=}B\overset{\diagup}{\diagdown}}$$

XXIV

Unsymmetrical diborylhydrazines having two borons attached to the same nitrogen of a hydrazine are not known. It is possible that steric factors inhibit the formation of such derivatives; on the other hand, the replacement of one hydrogen of a hydrazine moiety, $\overset{H}{\underset{H}{\diagup}}N-N\overset{\diagup}{\diagdown}$ by a boryl group naturally modifies the chemical reactivity of the remaining proton. However, hydrazinoboranes of type XXV are well established.

$$\overset{\diagdown}{\diagup}B-\overset{\cdot\cdot}{N}-\overset{\cdot\cdot}{N}\overset{\diagup}{\diagdown}$$
$$\underset{\text{XXV}}{\overset{|}{H}}$$

Although the first attempts to prepare hydrazinoboranes date back to EMELEUS[2] and SCHLESINGER[3], it was not until 1961 that H_2B-NH- $-NH-BH_2$ was obtained by pyrolysis of the diborane-hydrazine adduct[4]. Hydrazinoboranes, which are substituted at the boron atom by organic groups, are obtained in good yield through the reaction of tetraalkyl-diborane, $(R_2BH)_2$, with hydrazine at $100-150°$ [5].

$$(R_2BH)_2 + N_2H_4 \rightarrow R_2B-NH-NH-BR_2 + 2H_2 \qquad (II\text{-}52)$$

[1] NIEDENZU, K., P. FRITZ and J. W. DAWSON: Inorg. Chem. **3**, 778 (1964).
[2] EMELEUS, H. J., and F. G. A. STONE: J. chem. Soc. (London) **1951**, 840.
[3] STEINDLER, M. J., and H. I. SCHLESINGER: J. Amer. chem. Soc. **75**, 756 (1953).
[4] GOUBEAU, J., and E. RICHTER: Z. anorg. allg. Chem. **310**, 123 (1961).
[5] NÖTH, H.: Z. Naturforsch. **16b**, 471 (1961).

Adducts of the amine-borane type, $R_3B \cdot N_2H_4$, are the primary products obtained when a trialkylborane is reacted with hydrazine. These adducts decompose at temperatures above 170° and, in the presence of excess BR_3, symmetrical diborylhydrazines, XXIV, are formed. Hydrazinoboranes of type XXV are the primary products produced when a lesser amount of trialkylborane is employed; they subsequently disproportionate in an equilibrium reaction according to the equation:

$$2\,R_2B\text{—}NH\text{—}NH_2 \rightleftharpoons R_2B\text{—}NH\text{—}NH\text{—}BR_2 + N_2H_4 \qquad \text{(II-53)}$$

This disproportionation reaction is typical of those hydrazinoboranes having a terminal NH_2 group. Substitution of hydrogen by organic groups provides for stable hydrazinoboranes of type XXV.

The hydrazinolysis of diorganochloroboranes is not always unequivocal although R_2BCl and N,N-dimethylhydrazine do form $R_2B\text{—}NH\text{—}N(CH_3)_2$ in a smooth reaction. With free hydrazine, however, R_2BCl yields the salt-like compound XXVI as the main product[2].

$$\left[\begin{matrix} H_4N_2 \diagdown & \diagup R \\ & B \\ H_4N_2 \diagup & \diagdown R \end{matrix} \right]^{\oplus} Cl^{\ominus}$$

XXVI

The best yields of hydrazinoboranes have been obtained by the hydrazinolysis of an aminoborane[1, 2].

$$R_2B\text{—}NR_2 + N_2H_4 \rightleftharpoons R_2B\text{—}NH\text{—}NH_2 + R_2NH \qquad \text{(II-54)}$$

This reaction is also possible when using substituted hydrazines. Trisaminoboranes[2], bisaminoboranes[1] and alkylmercaptoboranes[3], R_2BSR', can be used in place of the monoaminoboranes. Whereas hydrazinolysis of bisaminoboranes with substituted hydrazines readily yields the bishydrazinoboranes, XXVII, the new ring system XXVIII is formed with free hydrazine[1].

$$\begin{matrix} HN\text{——}NR'_2 \\ \diagup \\ R\text{—}B \\ \diagdown \\ HN\text{——}NR'_2 \end{matrix} \qquad\qquad \begin{matrix} HN\text{——}NH \\ \diagup \quad\quad \diagdown \\ R\text{—}B \quad\quad B\text{—}R \\ \diagdown \quad\quad \diagup \\ HN\text{——}NH \end{matrix}$$

XXVII XXVIII

Derivatives of this ring system are described in more detail in Chapter IV-D.

Most hydrazinoboranes of the above linear types are monomeric. However, if steric and certain electronic factors permit, dimerization can occur. For example, (dimethylhydrazino)diphenylborane, XXIX, can be transhydrazinolized with free hydrazine (eq. II-55).

$$(C_6H_5)_2B\text{—}NH\text{—}N(CH_3)_2 + N_2H_4 \rightarrow (C_6H_5)_2B\text{—}NH\text{—}NH_2 + (CH_3)_2N\text{—}NH_2$$
$$\text{XXIX} \qquad\qquad\qquad \text{XXX} \qquad\qquad\qquad \text{(II-55)}$$

[1] NIEDENZU, K., H. BEYER and J. W. DAWSON: Inorg. Chem. 1, 738 (1962).
[2] NÖTH, H., and W. REGNET: Advances in Chemistry 42, 166 (1964).
[3] MIKHAILOV, B. M., and Y. A. BUBNOV: Proc. Acad. Sci. USSR., Div. Chem. Sci. 1960, 368.

The resultant (hydrazino)diphenylborane XXX is a dimer as indicated by molecular weight determinations: a cyclic structure, XXXI, has been proposed for this compound[1]. Such dimerization corresponds close-

$$\begin{array}{ccc} & \text{NH—NH}_2 & \\ (C_6H_5)_2B & & B(C_6H_5)_2 \\ & H_2N—NH & \\ & \text{XXXI} & \end{array}$$

ly to that of corresponding aminodiorganoboranes with a free NH_2 group. In general, all types of hydrazinoboranes are sensitive towards hydrolysis and oxidize quite readily; also they react with electrophilic reagents, such as HX or BX_3. The nature of the reaction intermediates has not yet been clearly established. However, in analogy to amino-boranes, the final products are formed through cleavage of the boron-nitrogen bond.

$$R_2B—NH—N\Big\langle \xrightarrow{\text{HX}} R_2BX + H_2N—N\Big\langle \qquad (\text{II-56})$$

The chemistry of azidoboranes has been explored only superficially WIBERG and MICHAUD[2] described the preparation of trisazidoborane, $B(N_3)_3$, and the lithium salt $Li[B(N_3)_4]$. On reacting dimesitylfluoro-borane with aluminium azide, a material was obtained which exhibited a strong infrared absorption near 2100 cm^{-1} [3]. Since this frequency had been assigned to an unsymmetrical vibration of a boron-bonded azido group, it was considered evidence for the formation of azididi-mesitylborane. On pyrolysis, rearrangement of the material occurred with migration of a mesityl group and the elimination of nitrogen as illustrated in the equation:

$$n\ R_2B—N_3 \rightarrow (—BR—NR—)_n + n\ N_2 \qquad (\text{II-57})$$

Detailed studies on some monoazidoboranes have been reported by PAETZOLD[3] who was able to obtain the explosive azidodichloroborane, Cl_2BN_3, by reacting stoichiometric amounts of trichloroborane with lithium azide. Pyrolysis of the Cl_2BN_3 at temperatures above 150° follows the pattern illustrated in eq. II-57 and yields hexachloroborazine, $(—BCl—NCl—)_3$. It is likely that the reaction proceeds via the inter-mediate Cl_2BN which then rearranges through migration of chlorine. This mechanism is substantiated by the observation that pyrolysis of pyridine-azidodiphenylborane in the presence of 2,5-diphenyltetrazole gives a small yield of 1,2,4,5-tetraphenyl-2,3,4,1-triazaboracyclopentene[4], indicating the existence of an intermediate $C_6H_5BNC_6H_5$. This postu-lation finds support in the formation of a dimer $(C_6H_5BNC_6H_5)_2$ from the pyrolysis of azidodiphenylborane. Since pyridine-azidodiphenyl-borane yields hexaphenylborazine in an analogous reaction, it seems

[1] NÖTH, H., and W. REGNET: Advances in Chemistry 42, 166 (1964).
[2] WIBERG, E., and H. MICHAUD: Z. Naturforsch. 9b, 497, 499 (1954).
[3] LEFFLER, J. E., and L. J. TODD: Chem. and Ind. 1961, 512.
[4] PAETZOLD, P. I.: Z. anorg. allg. Chem. 326, 47, 53, 58, 64 (1963).

reasonable to conclude that monoazidoboranes decompose in a stepwise mechanism as illustrated in eqs. II-58 to II-60, in which the final polymerization step appears to be influenced by steric factors.

$$R_2B—N_3 \rightarrow [R_2B—N] + N_2 \qquad \text{(II-58)}$$

$$[R_2B—N] \rightarrow [RB=NR] \qquad \text{(II-59)}$$

$$n[RB=NR] \rightarrow (—BR—NR—)_n \qquad \text{(II-60)}$$

Some (dialkylamino)azidoboranes, $R_2N—B(N_3)_2$ and $(R_2N)_2—B(N_3)$, have been obtained from reacting the corresponding (dialkylamino)chloroboranes with lithium azide[1]. Pyrolysis of bis(diethylamino)azidoborane results in migration of one dialkylamino group from boron to nitrogen to yield a polymeric compound $[R_2N—\overset{|}{B}—\overset{|}{N}—NR_2]_x$ with the elimination of nitrogen. This method of decomposition appears to be analogous to the mechanism described in eqs. II-58 to II-60.

a. Dimethylhydrazinodibutylborane[2]

A quantity, 16 g., of di-n-butylchloroborane is added with stirring to a solution of 12 g. of anhydrous dimethylhydrazine in 100 cc. of dry ether. After the exothermic reaction has subsided, the mixture is refluxed for one hour. The solid is filtered off and the filtrate stripped of ether. Distillation of the residue under reduced pressure affords 12 g. (77%) of the desired compound, b.p. 47—48 (2 mm.).

b. Bis(N,N-dimethylhydrazino)phenylborane[2]

A solution of 44 g. of bis(dimethylamino)phenylborane in 50 cc. of dry benzene is added slowly to a solution of 36 g. of anhydrous N,N-dimethylhydrazine in 100 cc. of benzene. The mixture is refluxed for one hour and the solvent and excess hydrazine are stripped off. The residue is distilled under vacuum to yield 39 g. of the desired compound, b.p. 80° (2 mm.).

c. Tris(N,N-dimethylhydrazino)borane[3]

A mixture of 14.2 g. of tris(dimethylamino)borane and 25 g. of anhydrous N,N-dimethylhydrazine was heated to 80° for several hours. The excess hydrazine was removed and the solid residue sublimed in vacuum. M.p. 104°.

H. Amino Derivatives of Diborane(4)

Derivatives of the type $R_2B—BR_2$ [4] with boron-boron bonding have, until recently, been prepared primarily through the use of tetrachlorodiborane(4), B_2Cl_4; both boron atoms of tetrachlorodiborane(4) act as acceptors. The compound can, therefore, add two molecules of a monodentate nitrogen donor by ordinary coordination. Such adducts of tetrachlorodiborane(4) have been described with trimethylamine[5, 6], dimethyl-

[1] PAETZOLD, P. I., and G. MAIER: Angew. Chem. 76, 343 (1964).

[2] NIEDENZU, K., H. BEYER and J. W. DAWSON: Inorg. Chem. 1, 738 (1962).

[3] NÖTH, H., and W. REGNET: Advances in Chemistry 42, 166 (1964).

[4] Chemical Abstracts has named diboron compounds as derivatives of the hypothetical $H_2B—BH_2$, diborane(4). Hence, a compound (RO)ClB—BCl(OR) is properly designated as 1,2-dialkoxy-1,2-dichlorodiborane(4).

[5] URRY, G., T. WARTIK, R. E. MOORE and H. I. SCHLESINGER: J. Amer. chem. Soc. 76, 5293 (1954).

[6] HOLLIDAY, A. K., F. J. MARSDEN and A. G. MASSEY: J. chem. Soc. (London) 1961, 3348.

amine[1], pyridine[2], hydrogen cyanide[2] and acetonitrile[2]. Also this subject has recently been reviewed[3]. Donor molecules which can eliminate hydrogen chloride, i.e. amines other than tertiary, react with tetrachlorodiborane(4) through aminolysis of the boron-halogen bond. This reaction has been reported for ammonia[4], hydrazine[2], ethylenediamine[2] and dimethylamine[2, 4]. Except in the case where dimethylamine is a reactant, the boron-containing products of the aminolysis reaction are nonvolatile solids of undefined structure. Dimethylamine, however, gives a mixture of tetrakis(dimethylamino)diborane(4), [(CH$_3$)$_2$N]$_2$B—B[N(CH$_3$)$_2$]$_2$, tris(dimethylamino)borane, B[N(CH$_3$)$_2$]$_3$, and some polymeric byproducts[2, 4]. Reaction of tetrachlorodiborane(4) with (dimethylamino)dimethylborane is reported to yield 1,2-bis(dimethylamino)-1,2-dichlorodiborane(4), XXXII, in polymeric form[5].

$$\begin{array}{cc} Cl & Cl \\ B\!\!-\!\!B & \\ (CH_3)_2N & N(CH_3)_2 \end{array}$$
XXXII

The veracity of this report appears to be somewhat in question in view of the fact that the cited compound has been obtained by other methods and does not seem inclined to polymerize. Instead, it tends to slowly decompose[6] even at room temperature.

Since 1960—61 a wide variety of amino derivatives of diborane(4) have been prepared by a more elegant method. Bisaminochloroboranes, (R$_2$N)$_2$BCl, were found to react with a finely dispersed alkali metal affording tetrakisaminodiborane(4) derivatives in high yield[7, 8].

$$2 \begin{array}{c} R_2N \\ \diagdown \\ B\!\!-\!\!Cl + 2\,Me \\ \diagup \\ R_2N \end{array} \longrightarrow \begin{array}{c} R_2N \diagdown \quad \diagup NR_2 \\ B\!\!-\!\!B \\ R_2N \diagup \quad \diagdown NR_2 \end{array} + 2\,MeCl \qquad (II\text{-}61)$$
XXXIII

These tetrakisaminodiboranes(4), XXXIII, are thermally quite stable in contrast to the tetraalkyldiboranes(4). Observations such as these suggest the existence of back-coordination of free electron pairs from nitrogen to boron. This assumption is supported by infrared spectroscopic studies[9]; ^{11}B nuclear magnetic resonance data are not yet available.

---

[1] APPLE, E. F., and T. WARTIK: J. Amer. chem. Soc. 80, 6158 (1958).

[2] HOLLIDAY, A. K., F. J. MARSDEN and A. G. MASSEY: J. chem. Soc. (London) 1961, 3348.

[3] HOLLIDAY, A. K., and A. G. MASSEY: Chem. Review 1962, 303.

[4] URRY, G., T. WARTIK, R. E. MOORE and H. I. SCHLESINGER: J. Amer. chem. Soc. 76, 5293 (1954).

[5] HOLLIDAY, A. K., A. G. MASSEY and F. B.TAYLOR: Proc. chem. Soc. (London) 1960, 359.

[6] NÖTH, H., and W. MEISTER: Z. Naturforsch. 17b, 714 (1962).

[7] BROTHERTON, R. J., A. L. McCLOSKEY, L. L. PETTERSON and H. STEINBERG: J. Amer. chem. Soc. 82, 6242 (1960).

[8] NÖTH, H., and W. MEISTER: Chem. Ber. 94, 509 (1961).

[9] BECHER, H. J., W. NOWODNY, H. NÖTH and W. MEISTER: Z. anorg. allg. Chem. 314, 226 (1962).

Derivatives of type XXXIV, with mixed substituents at the boron, have been prepared by a method analogous to that described in equation II-61. Such derivatives show a reasonable thermal

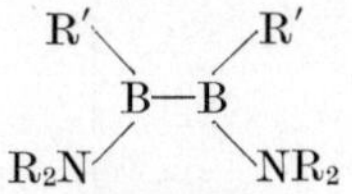

$$
\begin{array}{ccc}
R'\diagdown & & \diagup R' \\
& B{-}B & \qquad R, R' = alkyl,\ aryl \\
R_2N\diagup & & \diagdown NR_2 \\
& XXXIV
\end{array}
$$

stability[1, 2]. The chemistry of tetrakis(dimethylamino)diborane(4), which is most readily available by the method illustrated above, has been studied in some detail. The compound easily undergoes transamination with a variety of primary and secondary amines[3, 4]. The major limiting factor for this reaction appears to be the structure and size of the entering amine. Both amino groups of aliphatic or aromatic diamines have been found to attack the same boron atom in tetrakis-(dimethylamino)diborane(4) to afford bicyclic derivatives of type XXXV[4]. However, the use of 1,2-diaminodiboranes(4) such as those illustrated by XXXIV, provides for the incorporation of two directly

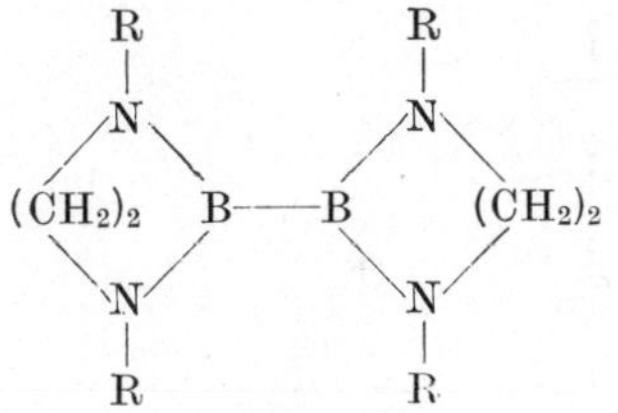

XXXV 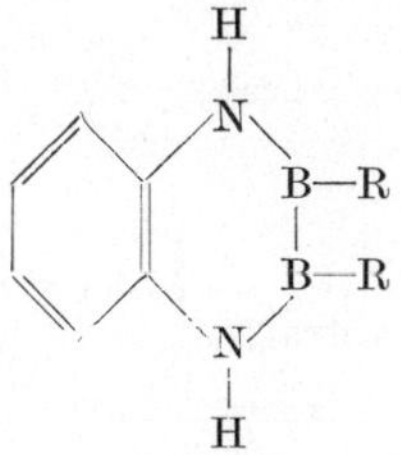   XXXVI

bonded boron atoms in the same ring, XXXVI[1]. Tetrakisaminodiboranes(4) are sensitive towards protolytic attack. In the presence of hydrogen chloride (to assist removal of the eliminated amine as a salt) alcoholysis of tetrakis(dimethylamino)diborane(4) provides tetraalkoxy derivatives in high yield[5, 6]. Phenols have been used in the same reaction in place of the alcohols. On hydrolysis of the alkoxy compounds, the acid $B_2(OH)_4$ is obtained.

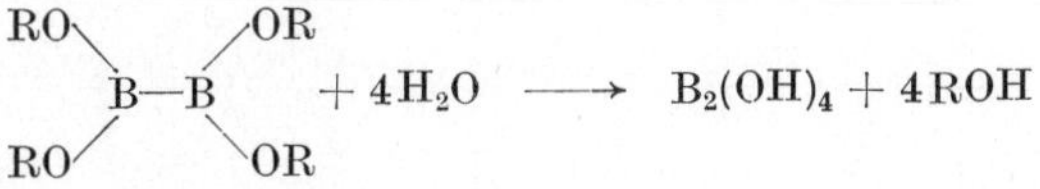

$$
\begin{array}{ccc}
RO\diagdown & \diagup OR \\
& B{-}B & + 4H_2O \longrightarrow B_2(OH)_4 + 4ROH \qquad (II\text{-}62) \\
RO\diagup & \diagdown OR
\end{array}
$$

[1] Nöth, H., and P. Fritz: Z. anorg. allg. Chem. **324**, 129 (1963).
[2] Brotherton, R. J., H. M. Manasevit and A. L. McCloskey: Inorg. Chem. **1**, 749 (1962).
[3] Brotherton, R. J., A. L. McCloskey, L. L. Petterson and H. Steinberg J. Amer. chem. Soc. **82**, 6242 (1960).
[4] Brown, M. P., A. E. Dann, D. W. Hunt and H. B. Silver: J. chem. Soc. (London) **1962**, 4648.
[5] Nöth, H., and W. Meister: Chem. Ber. **94**, 509 (1961).
[6] Brotherton, R. J., A. L. McCloskey, J. L. Boone and H. M. Manasevit: J. Amer. chem. Soc. **82**, 6245 (1960).

Table II-6. *Tetrakisaminodiboranes*(4)

	m.p., °C	b.p. (mm.) °C	Refer.
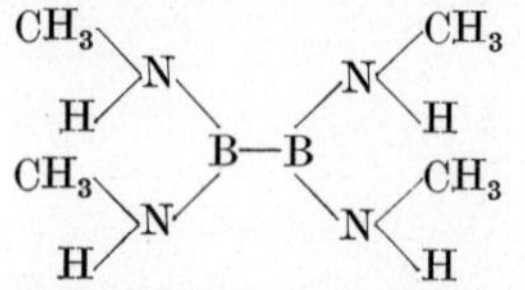 CH_3, CH_3 structure (see image)		25 (0.5)	1
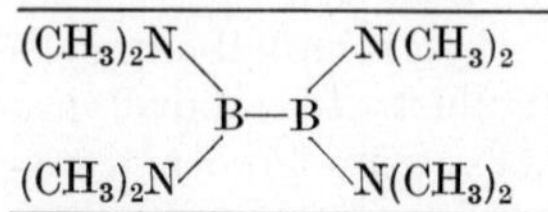 $(CH_3)_2N$, $N(CH_3)_2$ structure		55—57 (2)	1, 3
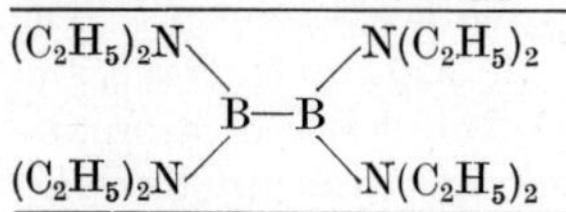 $(C_2H_5)_2N$, $N(C_2H_5)_2$ structure	−65	84 (0.1)	3
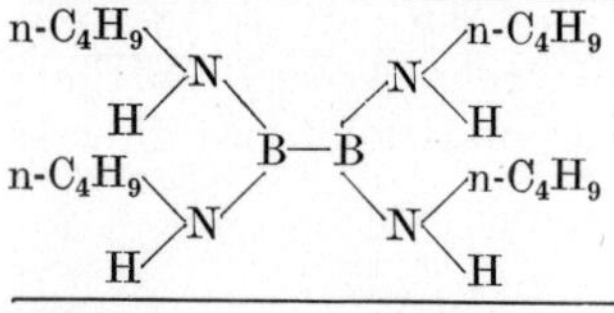 n-C_4H_9 structure		170—183 (0.55)	1
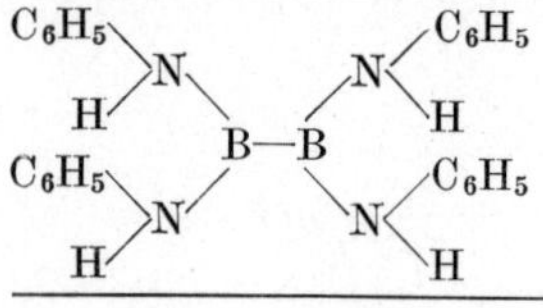 C_6H_5 structure	180—200		1
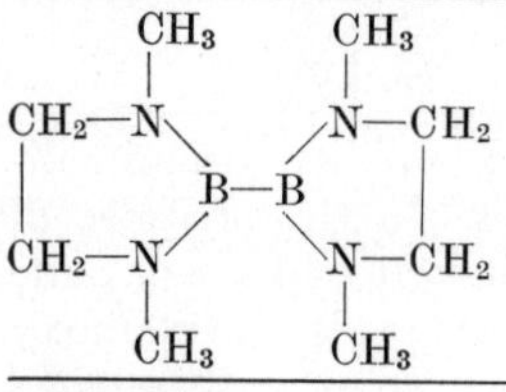 CH_3 / CH_3 ring structure	43—44	85 (5)	2
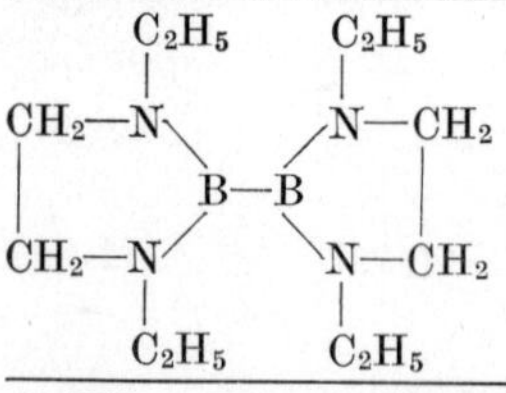 C_2H_5 / C_2H_5 ring structure		80 (0.1)	2
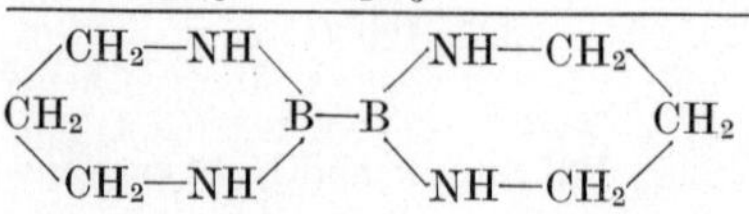 CH_2—NH ring structure		74—75 (0.6)	2

[1] BROTHERTON, R. J., A. L. McCLOSKEY, L. L. PETTERSON and H. STEINBERG: J. Amer. chem. Soc. **82**, 6242 (1960).

[2] BROWN, M. P., A. E. DANN, D. W. HUNT and H. B. SILVER: J. chem. Soc. (London) **1962**, 4648.

[3] NÖTH, H., and W. MEISTER: Chem. Ber. **94**, 509 (1961).

Thermal degradation of tetrakis(dimethylamino)diborane(4) at 300° results in substantial yields of bis(dimethylamino)borane, $[(CH_3)_2N]_2BH$, and tris(dimethylamino)borane, $B[N(CH_3)_2]_3$ [1]. In analogy, pyrolysis of 1,2-bis(dimethylamino)-1,2-dibutyldiborane(4), (XXXIV, $R = CH_3$, $R' = C_4H_9$) provides (dimethylamino)butylborane, $(CH_3)_2N\!-\!BHC_4H_9$; a polymeric dimethylaminoboron, $[(CH_3)_2NB]_x$ is also obtained as a by-product of this reaction along with butene which indicates migration of hydrogen from carbon to boron [3].

On reacting tetrakis(dimethylamino)diborane(4) with hydrogen chloride, cleavage of the boron-nitrogen bond occurs [3]. In vacuum and in the presence of excess hydrogen chloride, one obtains the dimethylamine adduct of tetrachlorodiborane(4).

$$B_2[N(CH_3)_2]_4 + 6\,HCl \rightarrow B_2Cl_4 \cdot 2\,HN(CH_3)_2 + 2[(CH_3)_2NH_2]Cl \quad (II\text{-}63)$$

In ethereal solution, however, it is possible to cleave only one or two of the B—N bonds to yield XXXVII and XXXVIII ($R{=}CH_3$). These two compounds are also obtained on reacting tetrakis(dimethylamino)di-

$$
(CH_3)_2N\diagdown\qquad\diagup Cl
$$
$$
B\!-\!B
$$
$$
(CH_3)_2N\diagup\qquad\diagdown N(CH_3)_2
$$
$$
\text{XXXVII}
$$

borane(4) with trichloroborane [4]. The reaction is easily controlled if this trihalogenoborane is added in the form of trimethylamine-trichloroborane [5]. Tribromoborane or methyldibromoborane react in an analogous

$$B_2(NR_2)_4 + BBr_3 \rightarrow B_2Br_2(NR_2)_2 + BrB(NR_2)_2 \quad (II\text{-}64)$$

manner [2]. The analogous reaction with trifluoroborane, however, is more complex; only nonvolatile solid materials are obtained and the reaction cannot be used for the preparation of tetrafluorodiborane(4) [6].

Dihalogeno derivatives of type XXXVIII are more reactive than the tetrakisaminodiboranes(4). This is illustrated, for example, by their ready hydrogenation:

$$
\begin{array}{ccc}
R_2N & NR_2 & \\
\diagdown & \diagup & \xrightarrow{H_2} \\
B\!-\!B & & \\
\diagup & \diagdown & \\
Cl & Cl & \\
\end{array}
\quad \left[\begin{array}{c} R_2N \\ \diagdown \\ B\!-\!H \\ \diagup \\ Cl \end{array}\right]_2 \quad (II\text{-}65)
$$
$$\text{XXXVIII}$$

1,2-Bis(dimethylamino)-1,2-diorganodiboranes(4), XXXIV, have been described previously; they are obtained by the alkali metal condensation

[1] PETTERSON, L. L., and R. J. BROTHERTON: Inorg. Chem. **2**, 423 (1963).
[2] NÖTH, H., and P. FRITZ: Z. anorg. allg. Chem. **324**, 129 (1963).
[3] NÖTH, H., and W. MEISTER: Z. Naturforsch. **17b**, 714 (1962).
[4] NÖTH, H., H. SCHICK and W. MEISTER: J. organometal. Chem. **1**, 401 (1964).
[5] BROWN, M. P., and H. B. SILVER: Chem. and Ind. **1963**, 85.
[6] BROTHERTON, R. J., A. L. MCCLOSKEY and H. M. MANASEVIT: Inorg. Chem. **2**, 41 (1963).

of (dimethylamino)alkylchloroboranes[1, 2], through the disproportionation
of tetrakis(dimethylamino)diborane(4) with trialkylboranes[3] or by the
alkylation of dihalogeno derivatives XXXVIII[4] using metalorganic
reagents. They react with hydrogen chloride at low temperatures
yielding a 1:2 adduct. In the presence of excess hydrogen chloride,
cleavage of the boron-boron bond results. This bond breakage also
occurs when these same compounds are attacked by halogens. When
trihalogenoboranes are used, cleavage of the boron-boron bond and
extensive disproportionation is observed, polymeric organoboron, $(RB)_n$,
being among the resultant products[1, 2].

The vibration spectrum of tetrakis(dimethylamino)diborane(4) was
reported by BECHER and coworkers[5]. Steric effects produce a structure
in which the NC_2 groups and the BN_2 groups are tilted against each
other. This results in a lowering of the force constant. Boron-nitrogen
stretching frequencies were observed near 1360 and 1414 cm.$^{-1}$ and a
valence force constant of about 5.5 to 6.0 mdyne/Å was assumed; the
B—B vibration was recorded at 586 cm.$^{-1}$.

Tetrakis(dimethylamino)diborane(4)[6]. —

A three-necked flask, equipped with stirrer, dropping funnel and reflux condenser is charged with a solution of 134 g. of bis(dimethylamino)chloroborane in
300 cc. of dry petrolether, b.p. 50—60°, under an inert atmosphere. Ten percent
of a potassium-sodium alloy made of 45 g. of potassium and 8 g. of sodium[7] is
added through a dropping funnel with vigorous stirring. When the reaction has
started[8], the remainder of the alloy is added over a period of one hour; the mixture
is then refluxed until the solution is free of chloride. The reaction mixture is cooled
to room temperature, filtered through a fritted disc in an inert gas atmosphere
and stripped of solvent. Distillation of the residue affords the desired compound
in a yield of about 90%, b.p. 86—87° (10 mm.).

---

[1] BROTHERTON, R. J., H. M. MANASEVIT and A. L. McCLOSKEY: Inorg. Chem.
1, 749 (1962).
[2] NÖTH, H., and P. FRITZ: Z. anorg. allg. Chem. 324, 129 (1963).
[3] MEISTER, W.: as cited in ref. (2).
[4] NÖTH, H., and W. MEISTER: Z. Naturforsch. 17 b, 714 (1962).
[5] BECHER, H. J., W. SAWODNY, H. NÖTH and W. MEISTER: Z. anorg. allg.
Chem. 314, 226 (1962).
[6] NÖTH, H., and W. MEISTER: Chem. Ber. 94, 509 (1961).
[7] To prepare the (liquid) potassium-sodium alloy, the two metals are kneaded
in a polyethylene beaker with a metal pestle under petrolether until all the metal
has liquified. A pipette, rinsed with petrolether, is slowly lowered through the
supernatant liquid to insure the inclusion of petrolether; the liquid metal alloy
is drawn into the pipette and transferred to the dropping funnel.
[8] Not more than 10% of the alloy should be added prior to the initiation of
the reaction. If the reaction does not start on stirring, gentle warming of the
reaction vessel is recommended.

Chapter III

The Borazines

A. Introduction

In 1926 STOCK and POHLAND[1] reacted diborane and ammonia and isolated a material analyzing as $B_3N_3H_6$. It was soon established that the molecule consisted of a ring structure of alternating boron and nitrogen atoms and ever since, borazine, $(—BH—NH—)_3$, and its derivatives have attracted the interest of the chemist primarily because of the resemblance of the six-membered boron-nitrogen heterocycle to benzene and its derivatives. Indeed, a comparison of a variety of physical constants of benzene and the parent borazine shows striking similarities.

Table III-1. *Physical Constants*

Constants	Borazine	Benzene
molecular weight	80.5	78.1
b.p., °K	328	353
m.p., °K	216	279
critical temperature, °K	525	561
liquid density at b.p., g./cm.³	0.81	0.81
crystal density at m.p.	1.00	1.01
surface tension at m.p., dynes/cm.	31.1	31.0
Trouton constant	21.4	21.1
parachor	208	206

Thus it is not surprising that borazine has been called "inorganic benzene" and that KEKULE structures (Ia—c) were proposed for the molecule.

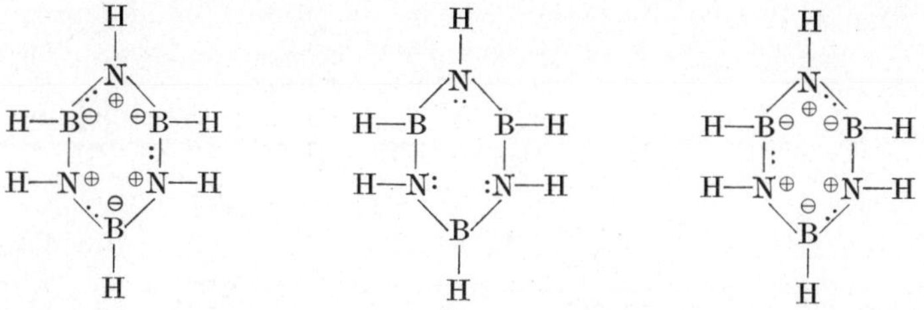

Ia   Ib   Ic

In structure Ia and Ic, the free electron pair of the nitrogen participates in the B—N bond, thus producing a formal negative charge at the boron atoms and a positive charge at the nitrogen atoms. It is therefore

---

[1] STOCK, A., and E. POHLAND: Ber. dtsch. chem. Ges. **59**, 2215 (1926).

evident that borazine can only formally be compared with benzene and today there is little doubt left that the aromaticity of this inorganic boron-nitrogen heterocycle has been overemphasized. Although even the bond distances in borazines are comparable to those of benzene, one should recognize that six-membered heterocyclic systems are quite common in inorganic chemistry, as evidenced by the trimeric phosphazenes, the thiazines, phosphinoboranes, boroxines, sulfur trioxide and others.

Table III-2

Bond Distances (in Å)

Borazine	Benzene
B—N 1.44	C—C 1.42
N—H 1.02	C—H 1.08
B—H 1.20	

Early work in borazine chemistry was severely hampered by the difficulties in synthesizing and handling the materials. These obstacles have since been overcome as is illustrated by the fact that some ten years ago only about two dozens of substituted borazines were known whereas today their number approaches three hundred. Particularly since LAUBENGAYER introduced his now classical synthesis of B-trichloroborazine[1], the practice of using high vacuum techniques has been partially supplanted by more versatile approaches and large-scale preparations can presently be effected with ordinary laboratory equipment. The utilization of spectroscopy as a tool for explaining both structures and mechanisms has been stimulated and is beginning to shed more light on the unique nature of the borazine ring.

B. The Parent Borazine, (—BH—NH—)₃

1. Preparation

The basic method for the synthesis of the borazine ring involves the condensation of boron hydrides with amines. The parent compound, $(—BH—NH—)_3$, was first prepared in this manner and was the first recognized borazine. It was obtained by STOCK and POHLAND in low yield by the thermal decomposition of the adduct of ammonia with diborane at 200° in a bomb tube.[2] This preparation formally follows the scheme:

$$BH_3 \xrightarrow{NH_3} H_3B \cdot NH_3 \xrightarrow{-H_2} H_2B—NH_2 \xrightarrow{-H_2} HB{=}NH \xrightarrow{cyclisation} (—BH—NH—)_3$$

$$(III\text{-}1)$$

WIBERG and BOLZ[3] found that the yield of borazine increases with the reaction temperature. The reaction time does not appear to effect the yield substantially but does influence the hydrogen content of the solid byproducts. Depending upon the reaction conditions, a certain amount of $H_2B—NH_2$ intermediate polymerizes irreversibly to form polymeric

[1] BROWN, C. A., and A. W. LAUBENGAYER: J. Amer. chem. Soc. 77, 3699 (1955).
[2] STOCK, A., and E. POHLAND: Ber. dtsch. chem. Ges. 59, 2215 (1926).
[3] WIBERG, E., and A. BOLZ: ibid. 73, 209 (1940).

(BH—NH)$_x$ and (BNH)$_x$. Therefore, rapid heating of the diborane-
ammonia adduct to high temperatures inadvertently favors the formation
of borazine[1]. Reports on the influence of pressure in this reaction are
not clear. STOCK and POHLAND[2] and WIBERG and BOLZ[3] worked preferentially at pressures below one atmosphere and were able to obtain
a yield of about 45%; the yield decreased to 23% as the pressure was
raised to five atmospheres. SCHLESINGER and coworkers[4, 5] reported
increasing yields with increasing press-
ure as illustrated in Table III-3. An
excess of ammonia decreases the yield
of borazine whereas, in the presence of
excess diborane, increasing amounts of
aminodiborane, B$_2$H$_5$NH$_2$, are formed[3].
After years of experience, the best yields
of approximately 50% were obtained by
rapidly heating B$_2$H$_6$ and NH$_3$ in a care-
fully adjusted molar ratio of 1:2 to about 250° at one atmosphere for
around 45 minutes.

Table III-3. *Influence of Pressure
on Yield of Borazine at 200°*

pressure, atm.	yield, %
1.8	27
4.1	33
4.8	38
11.0	41.5

It is possible that the formation of borazine by this reaction proceeds
via the cation [NH$_3$BH$_2$NH$_3$]$^\oplus$ [6]. This presumption is substantiated
by the fact that ammonia adducts of higher boron hydrides which are
also known to contain this ion, similarly produce borazine on pyrolysis[2, 7–9].

The use of this basic condensation reaction for the synthesis of
borazine was severely hampered by the difficulty encountered in preparing and handling the boron hydrides. However, more convenient
preparative procedures have since been developed. First, metal hydroborates were introduced as a readily available source of boron hydride,
ammonia was replaced by ammonium chloride and the reaction was
performed in anhydrous ethereal solvents [10-13]. A major breakthrough
in preparative technique was achieved by BROWN and LAUBENGAYER[14]

[1] WIBERG, E., K. HERTWIG and A. BOLZ: Z. anorg. allg. Chem. **256,** 177 (1948).
[2] STOCK, A., and E. POHLAND: Ber. dtsch. chem. Ges. **59,** 2215 (1926).
[3] WIBERG, E., and A. BOLZ: ibid. **73,** 209 (1940).
[4] SCHLESINGER, H. I., L. HORVITZ and A. B. BURG: J. Amer. chem. Soc. **58,**
409 (1936).
[5] SCHLESINGER, H. I., D. M. RITTER and A. B. BURG: ibid. **60,** 1296 (1938).
[6] This would be analogous, for instance, to the formation of halogenophos-
phazenes, (PNX$_2$)$_n$, through the thermal decomposition of PCl$_5$/NH$_4$Cl which
reportedly occurs via similar ions[15].
[7] SCHLESINGER, H. I., D. M. RITTER and A. B. BURG: J. Amer. chem. Soc.
60, 2297 (1938).
[8] STOCK, A., E. WIBERG and H. MARTINI: Ber. dtsch. chem. Ges. **63,** 2927 (1930).
[9] STOCK, A., and E. POHLAND: ibid. **62,** 90 (1929).
[10] SCHAEFFER, G. W., and E. R. ANDERSON: J. Amer. chem. Soc. **71,** 2143 (1949).
[11] SCHAEFFER, G. W., R. SCHAEFFER and H. I. SCHLESINGER: ibid. **73,** 1612
(1951).
[12] MIKHEEVA, V. I., and V. Y. MARKINA: J. inorg. Chem. USSR. **1,** 2200 (1956).
[13] HAWORTH, D. T., and L. F. HOHNSTEDT: Chem. and Ind. **1960,** 559.
[14] BROWN, C. A., and A. W. LAUBENGAYER: J. Amer. chem. Soc. **77,** 3699
(1955).
[15] BECKE-GOEHRING, M., and W. LEHR: Z. anorg. allg. Chem. **327,** 128 (1964).

who reacted ammonium chloride with trichloroborane in chlorobenzene solution to afford B-trichloroborazine.

$$3\,BCl_3 + 3\,NH_4Cl \rightarrow (-BCl-NH-)_3 + 9\,HCl \qquad (III\text{-}2)$$

B-Trichloroborazine can readily be reduced to the parent borazine with either lithium hydroborate[1, 2], or, even better, with sodium hydroborate[3, 4]. This latter method today appears to be the most efficient and convenient procedure for preparing laboratory quantities of borazine.

Borazine by Reduction of B-trichloroborazine[4]

A two-liter, three-necked flask equipped with stirrer, reflux condenser and dropping funnel is charged with 77 g. of sodium hydroborate dissolved in about 600 cc. of anhydrous diglyme. The flask is chilled to 0° and 380 g. (about 25% excess) of tri-n-butylamine is added with vigorous stirring. The reaction flask is flushed with dry nitrogen and a solution of 100 g. of B-trichloroborazine in 250 cc. of diglyme is added dropwise over a one hour period. The stirring is continued for about thirty minutes after the addition is complete. The reaction flask is now connected to a spiral reflux condenser, cooled with a salt-ice bath, and the reaction mixture is distilled in an oil pump vacuum, first at room temperature, later with gentle heating to 40—50°. Borazine and some of the ether collect in a Dry Ice-cooled trap. The crude borazine is then transferred in an inert atmosphere (argon or nitrogen) to a PODBIELNIAK distillation column which has been thoroughly flushed with the inert gas. The cold finger is cooled with circulating brine and the mixture is distilled at atmospheric pressure, while the receiving flask is cooled with an ice bath. Yield: 20.3 g. (46%). Note: The dimethyl ether of diethylene glycol (diglyme) must be absolutely dry. This is achieved best by refluxing it over calcium hydride and subsequently distilling it over lithium aluminium hydride. Commercial tri-n-butylamine is refluxed with acetic acid and distilled at atmospheric pressure. In place of the PODBIELNIAK distillation a 25 cm. silver mantle column charged with stainless steel helices has been used successfully.

2. Properties and Structure of Borazine

Borazine is a clear colorless liquid, m.p. −58°, b.p. 55°. The molecular composition $B_3N_3H_6$ was determined by elemental analysis and quantitative hydrolysis studies. In the gas phase, borazine is thermally quite stable; this fact militates against the existence of boron-boron bonds within the molecule. Electron diffraction and X-ray data[5, 6] are consistent with a planar hexagonal molecule of alternating B—H and N—H groups as shown in structure I, and the B—N bond distance is intermediate between that expected for a single bond (1.54 Å) and that for a double bond (1.36 Å). RAMASWAMY[7] found a dipole moment of 0.67 Debye for borazine and SHELDON and SMITH suggested[8] that this moment

[1] SCHAEFFER, R., M. STEINDLER, L. F. HOHNSTEDT, H. S. SMITH JR., L. B. EDDY and H. I. SCHLESINGER: J. Amer. chem. Soc. **76**, 3303 (1954).
[2] EDDY, L. B., H. S. SMITH JR. and R. R. MILLER: ibid. **77**, 2105 (1955).
[3] HOHNSTEDT, L. F., and D. T. HAWORTH: ibid. **82**, 89 (1960).
[4] DAHL, G. H., and R. SCHAEFFER: J. inorg. nucl. Chem. **12**, 380 (1960).
[5] STOCK, A., and R. WIERL: Z. anorg. allg. Chem. **203**, 228 (1931).
[6] BAUER, S. H.: J. Amer. chem. Soc. **60**, 524 (1938).
[7] RAMASWAMY, K. L.: Proc. Indian Acad. Sci. **2**A, 364, 630 (1935).
[8] SHELDON, J. C., and B. C. SMITH: Chem. Revs. **1960**, 200.

was caused by the presence of impurities. However, the dielectric constants and densities of dilute solutions of borazine in benzene were measured by WATANABE and KUBO[1] and were used to calculate a dipole moment of 0.50 Debye. This small dipole moment might suggest a minor deviation from the coplanar structure. This occurrence, however, is not in agreement with the crystallographic data. It seems more reasonable to postulate the presence of a large atom polarization in the molecule.

RAMAN and infrared spectra of borazine have been described in great detail[2, 3]. Assignment of fundamental frequencies on the basis of benzene-like symmetry D$_{3h}$ gave satisfactory correlations. Certain discrepencies between observed and calculated values for the out-of-plane hydrogen vibrations were re-investigated by SPURR and CHANG[4]. By use of a simple potential function for borazine, these authors obtained a more satisfactory agreement with the experimental data in their calculations. However, the assignment of B—H and N—H bending modes still appears to be controversial. From the shape and intensity of the absorption curves of N-trideuteroborazine and B-trichloro-N-trideuteroborazine, KUBO and coworkers[5] concluded that planar N—H bending does not have a higher frequency than the planar B—H bending. The assignments of fundamentals of borazine by various authors are illustrated in Table III-4.

Table III-4. *Assignments of Fundamentals of Borazine*

species, fundamentals		wavenumber (cm.$^{-1}$)	assignments		
			CRAWFORD and EDSALL[2]	PRICE et al.[3]	WATANABE et al.[5]
E'	ν_{11}	3490		NH stretching	
	ν_{12}	2530		BH stretching	
	ν_{13}	1605		BN stretching	
	ν_{14}	1465		BN stretching	
	ν_{15}	918	NH bending	NH, BH	BH bending
	ν_{16}	718	BH bending	bending	NH bending
	ν_{17}	519			ring distortion
A_2''	ν_8	1088	NH bending		BH bending
	ν_9	649	BH bending		NH bending
	ν_{10}	415		BN torsion	
A_1'	ν_1	3450		NH stretching	
	ν_2	2535		BH stretching	
	ν_3	938			ring distortion
	ν_4	851			BN stretching
E''	ν_{18}	1070	NH bending		BH bending
	ν_{19}	798	BH bending		NH bending
	ν_{20}	288		BN torsion	

[1] WATANABE, H., and M. KUBO: J. Amer. chem. Soc. **82,** 2428 (1960).

[2] CRAWFORD, B. L., and J. T. EDSALL: J. chem. Physics **7,** 223 (1939).

[3] PRICE, W. C., R. D. B. FRASER, T. S. ROBINSON and H. C. LONGUET-HIGGINS: Disc. Faraday Soc. **9,** 131 (1950).

[4] SPURR, R. A., and S. CHANG: J. chem. Physics **19,** 528 (1951).

[5] WATANABE, H., T. TOTANI, T. NAKAGAWA and M. KUBO: Spectrochim. Acta **16,** 1076 (1960).

Typical for all borazines is the observation of a very strong B—N stretching frequency in the infrared spectrum. It is easily recognized, since it is usually the strongest band of the spectrum and is normally accompanied by either a shoulder or a weaker band at slightly higher frequencies (about $10-15$ cm.$^{-1}$) denoting the ^{10}B isotope effect. Simultaneous observation of the B—N stretch in the 1400 to 1500 cm.$^{-1}$ region and a medium intensity B—N out-of-plane vibration near 700 cm.$^{-1}$ seems to be typical of a borazine ring and can be used for general diagnostic purposes in borazine chemistry[1].

The ultraviolet spectrum of borazine has been related to that of benzene[2, 3] and indeed both spectra show some similarities. A group of maxima between 1995 and 1895 Å in the borazine spectrum seems to correspond to the benzene bands near 2080 Å although the former are much weaker.

A simple semiempirical molecular orbital treatment (HÜCKEL method) on borazine was first reported by ROOTHAAN and MULLIKEN[4]. The applied model, however, required potentials of a neutral molecule, in contrast to the popular formulations as indicated in structures Ia and Ic in which the borazine nucleus consists of B$^\ominus$ and N$^\oplus$ atoms. Later the PARISER-PARR and POPLE methods for calculations of the electronic structures of π-systems based on a many-electron Hamiltonian were used successfully for the determination of organic π-electron systems. The first such application to inorganic systems was reported by DAVIES[5] who calculated the electronic spectrum of borazine by the self-consistent molecular orbital method based on the normal B$^\ominus$—N$^\oplus$ model of borazine. However, no definite conclusions with respect to absolute values were drawn. On the basis of measurements of the diamagnetic susceptibility, the anisotropy of a borazine ring was estimated by WATANABE, ITO and KUBO[6] to be about -36×10^{-6}. A molecular orbital treatment was then effected to evaluate the diamagnetic anisotropy as a function of molecular parameters. In agreement with the work by SPURR and CHANG[7], 24% of the classical double bond structure was derived, corresponding to a π-electron bond order of approximately 45% for the boron-nitrogen linkage. Recently, a more extensive molecular orbital treatment of borazine was performed by CHALVET, DAUDEL and KAUFMAN[8] on the basis of both models mentioned above. Surprisingly, the calculations based on the neutral model agreed more favorably when compared with the experimental and theoretical criteria (see also Chapter III-C 3).

[1] BEYER, H., J. B. HYNES, H. JENNE and K. NIEDENZU: Advances in Chemistry 42, 266 (1964).

[2] PLATT, J. R., H. B. KLEVENS and G. W. SCHAEFFER: J. chem. Physics 15, 598 (1947).

[3] JACOBS, L. E., J. R. PLATT and G. W. SCHAEFFER: ibid. 16, 116 (1948).

[4] ROOTHAAN, C. C. J., and R. S. MULLIKEN: ibid. 16, 118 (1948).

[5] DAVIES, D. W.: Trans. Faraday Soc. 56, 1713 (1960).

[6] WATANABE, H., K. ITO and M. KUBO: J. Amer. chem. Soc. 82, 3294 (1960).

[7] SPURR, R. A., and S. CHANG: J. chem. Physics 19, 528 (1951).

[8] CHALVET, O., R. DAUDEL and J. J. KAUFMAN: Advances in Chemistry 42, 251 (1964).

The proton magnetic resonance spectrum of borazine showed a ^{14}NH-triplet, a ^{11}BH quartet and part of a ^{10}BH septet[1, 2]. The broadening of multiplet components was explained in terms of a quadrupole relaxation of nuclei bonded to hydrogen. Chemical shifts of the B—H and N—H protons were interpreted on the grounds that the nitrogen atoms in borazine assume sp^2 hybridization and the π-electrons are presumed to migrate to some degree from nitrogen to boron.

A more recent theoretical consideration of the proton magnetic resonance spectrum of borazine has been advanced by SUZUKI and KUBO[3].

The entropy, enthalpy and free energy of gaseous borazine were calculated from spectral data[4] and vapor pressure data were utilized to obtain thermodynamic values for the liquid. A heat capacity equation for gaseous borazine was derived from these values. The heat of combustion was calculated to be: $\Delta H°(25° C) = -552.9 \pm 3.0$ kcal./mole   (III-3) from data obtained by burning liquid borazine in oxygen in a bomb calorimeter. From this value, the heat of formation was calculated to be -131.1 ± 3.2 kcal./mole for liquid borazine, and -124.1 ± 3.2 kcal./mole for gaseous borazine[5].

3. Chemical Reactions of Borazine

On prolonged storage of liquid borazine at room temperature, explosions without apparent cause have very often been observed. It is possible that slight amounts of impurities catalyze the decomposition which can also be initiated through the influence of light. MAMANTOV and MARGRAVE[6] investigated the decomposition of liquid borazine at ambient temperatures and reported the formation of more highly aggregated species which presumably are formed via B-aminoborazines, especially 2,4-diaminoborazine. On pyrolysis of gaseous borazine at temperatures above 340°, the formation of higher boron-nitrogen ring compounds was also observed; structures analogous to naphthalene (II) and biphenyl (III) have been found as well[7].

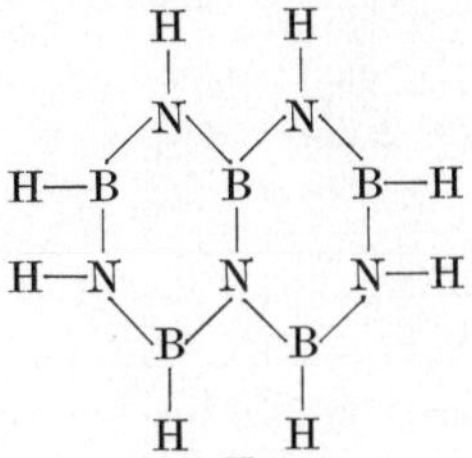

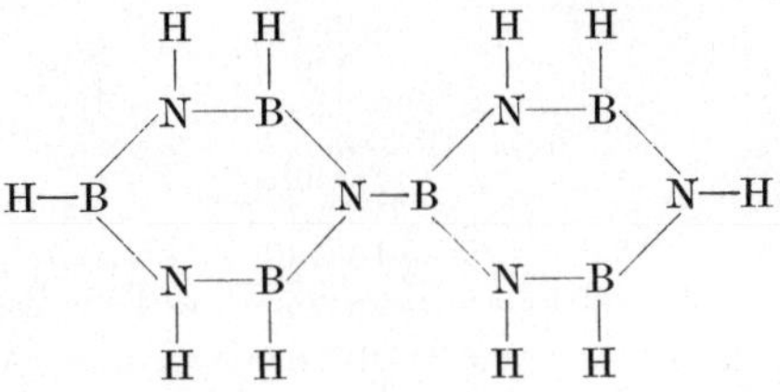

[1] ITO, K., H. WATANABE and M. KUBO: J. chem. Physics **32**, 947 (1959).

[2] ITO, K., H. WATANABE and M. KUBO: Bull. chem. Soc. Japan **33**, 1580 (1960).

[3] SUZUKI, M., and R. KUBO: Molecular Physics **7**, 201 (1964).

[4] CRAWFORD, B. L., and J. T. EDSALL: J. chem. Physics **7**, 223 (1939).

[5] KILDAY, M. V., W. H. JOHNSON and E. J. PROSEN: J. Res. Nat. Bur. Standards **65 A**, 101 (1961).

[6] MAMANTOV, G., and J. L. MARGRAVE: J. inorg. nucl. Chem. **20**, 348 (1961).

[7] LAUBENGAYER, A. W., P. C. MOEWS JR. and R. F. PORTER: J. Amer. chem. Soc. **83**, 1337 (1961).

Evidence for the formation of 2,4-diaminoborazine in this process indicated the breakdown of the molecule prior to the rearrangement which produces polynucleated borazines. In this connection, it is of interest to note that in the synthesis of B-trichloroborazine from trichloroborane and ammonium chloride as reported by BROWN and LAUBENGAYER[1], considerable amounts of nonvolatile materials are formed. In spite of the lack of confirming results, these might very well comprise similar polynucleated chlorine-containing borazines.

Borazine does not react with oxygen at room temperature, though it will explode with oxygen in an electric arc. It dissolves in water with the slow evolution of hydrogen and ammonia, indicating hydrolytic attack. The formation of B-trihydroxyborazine, $(-BOH-NH-)_3$, has been claimed[2] but this compound was never isolated and characterized. At room temperature, borazine reacts with methanol to give a 1:3 adduct. Pyrolysis of the adduct yields B-trimethoxyborazine, $(-BOCH_3-NH-)_3$, with the elimination of hydrogen[3]. Besides a nonvolatile, white solid, only ammonia-trimethoxyborane, $H_3N \cdot B(OCH_3)_3$, has been observed as a by-product. However, even at temperatures of $-30°$, ethanol produces alcoholysis of the boron-nitrogen bonds to afford mainly ammonia and triethoxyborane; no B-triethoxyborazine has been isolated from this reaction.

At lower temperatures, borazine adds chlorine; at room temperature this product evolves hydrogen chloride and an unidentified volatile material is obtained. At $0°$, two moles of bromine add to borazine and the adduct can decompose with the loss of two moles of hydrogen bromide; 2,4-dibromoborazine (IV) was isolated and identified as the primary product[2].

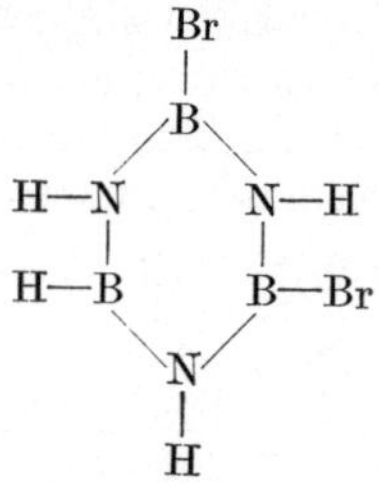

The interaction of borazine with hydrogen halides has been reported to yield B-trihalogenoborazines via 1:3 adducts, $B_3N_3H_6 \cdot 3HX$ [4]; however, these results could not be confirmed. A detailed study of this reaction by LAUBENGAYER and coworkers[5] confirmed the formation of an adduct $B_3N_3H_6 \cdot 3HX$, which at higher temperatures lost HX; the starting materials were recovered in substantial yield. In contrast to previous claims[2], no formation of B-trihalogenoborazine was observed. On treat-

[1] BROWN, C. A., and A. W. LAUBENGAYER: J. Amer. chem. Soc. 77, 3699 (1955).
[2] WIBERG, E., and A. BOLZ: Ber. dtsch. chem. Ges. 73, 209 (1940).
[3] HAWORTH, D. T., and L. F. HOHNSTEDT: J. Amer. chem. Soc. 81, 842 (1959).
[4] WIBERG, E.: Naturwissenschaften 35, 182, 212 (1948).
[5] LAUBENGAYER, A. W.: Personal communication.

ment of borazine with trihalogenoboranes, exchange of the B-attached hydrogen by halogen has been reported and moderate yields of mono- and dihalogenated borazine were obtained[1]. With trimethylamine, borazine forms a 1:1 addition complex. On pyrolysis of this complex, the starting materials are reformed along with trimethylamine-borane, $(CH_3)_3N \cdot BH_3$, and some hydrogen as the only volatile products[2]. Borazine reacts with phenylmagnesium bromide to provide B-arylation; B-mono-, di- and triphenylborazine were isolated[3].

DAHL and SCHAEFFER[4] studied a number of exchange reactions between borazine and deuterated compounds. On treatment of borazine with ND_3DCl and DCN, exchange of the nitrogen-attached hydrogen occurred at a rate comparable to that of addition. Exchange of the B-hydrogens occurred with D_2, $NaBD_4$ and deuterated diborane. Although borazine reacted rapidly with deuterated water and ethanol, no isotopically substituted borazine was detected in the recovered material.

C. Symmetrically Substituted Borazines (—BR—NR'—)$_3$

1. Formation of Substituted Borazines by Condensation Reactions

a. Condensation of Borohydrides and Amines

In 1938 SCHLESINGER, RITTER and BURG[5] investigated the interaction of diborane with methylamine and were able to isolate N-trimethyl-borazine, (—BH—NCH$_3$—)$_3$. Subsequent investigations of this same reaction[6, 7], utilizing the metal hydroborates then available, demonstrated that the reaction of lithium hydroborate with alkylammonium salts proceeds through the following steps:

$$3\,LiBH_4 + 3\,[RNH_3]Cl \xrightarrow{25°} 3\,H_2B—NHR + 6\,H_2 + 3\,LiCl \xrightarrow{250°}$$
$$(—BH—NR—)_3 + 3\,H_2 \tag{III-4}$$

The formation of N-substituted borazines can be accomplished in one step if the components are heated in a high-boiling ether as a solvent. Even better results have been obtained on reacting sodium hydroborate with primary amines[8]. In an extension of this concept, EMELEUS

[1] SCHAEFFER, G. W., R. SCHAEFFER and H. I. SCHLESINGER: J. Amer. chem. Soc. **73**, 1612 (1951).

[2] WEIBRECHT, W. E., and A. W. LAUBENGAYER: Abstr. of Papers, 145th National Meeting of the American Chemical Society, New York 1963, p. 4-N.

[3] MOEWS, P. C. JR., and A. W. LAUBENGAYER: Inorg. Chem. **2**, 1072 (1963).

[4] DAHL, G. H., and R. SCHAEFFER: J. Amer. chem. Soc. **83**, 3034 (1961).

[5] SCHLESINGER, H. I., D. M. RITTER and A. B. BURG: ibid. **60**, 1296 (1938).

[6] SCHAEFFER, G. W., and E. R. ANDERSON: ibid. **71**, 2143 (1949).

[7] RECTOR, C. W., G. W. SCHAEFFER and J. R. PLATT: J. chem. Physics **17**, 460 (1949).

[8] HAWORTH, D. T., and L. F. HOHNSTEDT: Chem. and Ind. **1960**, 559.

and WADE[1] investigated the reaction of diborane with alkyl nitriles and were able to obtain N-alkylated borazines:

$$3 B_2H_6 + 6 CH_3CN \rightarrow 2(-BH-NC_2H_5-)_3 \qquad (III\text{-}5)$$

Mass spectroscopic investigation of the products obtained from the interaction of diborane with N_2F_4 indicates the formation of fluorinated borazines[2] although no definite products were isolated. Diborane reacts smoothly with halogenated alkylamines to form addition products which, on thermal decomposition, yield the corresponding borazines[3]. This reaction has been used to prepare N-fluoroalkylborazines (eq. III-6).

$$3 B_2H_6 + 6 CF_3CH_2NH_2 \rightarrow 2(-BH-NCH_2CF_3-)_3 + 12 H_2 \qquad (III\text{-}6)$$

b. Reaction of Organoboranes with Ammonia and Amines

SCHLESINGER, HORVITZ and BURG[4] obtained B-methylated borazines on reacting methyldiborane with ammonia; with methylamine, they obtained materials methylated on both the boron and the nitrogen atoms. Unsymmetrically substituted borazines were formed as by-products in both reactions. This basic reaction has been used for the preparation of a variety of substituted borazines. WIBERG and HERTWIG[5] investigated the condensation of trimethylborane with methylamine to yield B-trimethyl-N-trimethylborazine; on reacting trimethylborane with ammonia, B-trimethylborazine was obtained[6]. Analogously, the thermal decomposition of the triethylborane-ammonia adduct yielded B-triethylborazine[7]; B-trimethyl-N-triphenylborazine was formed from the trimethylborane-aniline adduct[8]. Treatment of trialkylamine-alkylboranes with ammonia and a trace of ammonium chloride as a catalyst in diglyme solution at 100—150°, results in a rapid evolution of hydrogen and a nearly quantitative yield of the corresponding B-trialkylborazine[9, 10].

$$RH_2B \cdot N(CH_3)_3 + NH_3 \rightarrow RH_2B \cdot NH_3 + N(CH_3)_3 \qquad (III\text{-}7)$$
$$3 RH_2B \cdot NH_3 \rightarrow (-BR-NH-)_3 + 6 H_2 \qquad (III\text{-}8)$$

The exact role of the ammonium chloride catalyst in this reaction is still unknown. However, it is apparent that the ammonium ion may function as a proton source under the conditions of reaction and thus catalyze the removal of hydridic hydrogen from an intermediate amine-

[1] EMELEUS, H. J., and K. WADE: J. chem. Soc. (London) **1960**, 2614.
[2] PEARSON, R. K., and J. W. FRAZER: J. inorg. nucl. Chem. **21**, 188 (1961).
[3] LEFFLER, A. J.: Inorg. Chem. **3**, 145 (1964).
[4] SCHLESINGER, H. I., L. HORVITZ and A. B. BURG: J. Amer. chem. Soc. **58**, 409 (1936).
[5] WIBERG, E., and K. HERTWIG: Z. anorg. allg. Chem. **255**, 141 (1947).
[6] WIBERG, E., K. HERTWIG and A. BOLZ: ibid. **256**, 177 (1948).
[7] ZHIGACH, A. F., E. B. KAZAKOVA and E. S. KRONGAUZ: Proc. Acad. Sci. USSR., Div. Chem. Sci. **1956**, 1029.
[8] BECHER, H. J., and S. FRICK: Z. anorg. allg. Chem. **295**, 83 (1958).
[9] HAWTHORNE, M. F.: J. Amer. chem. Soc. **81**, 5836 (1959).
[10] HAWTHORNE, M. F.: ibid. **83**, 833 (1961).

borane. An example of such a reaction sequence is illustrated by the following equations;

$$RH_2B \cdot NH_3 + NH_4^{\oplus} \rightarrow H_2 + NH_3 + RHB \cdot \overset{\oplus}{N}H_3 \qquad (III-9)$$

$$RHB \cdot \overset{\oplus}{N}H_3 + NH_3 \rightarrow RHB—NH_2 + NH_4^{\oplus} \qquad (III-10)$$

$$RHB—NH_2 + NH_4^{\oplus} \rightarrow RB=\overset{\oplus}{N}H_2 + NH_3 + H_2 \qquad (III-11)$$

$$RB=\overset{\oplus}{N}H_2 + NH_3 \rightarrow \frac{1}{3}(—BR—NH—)_3 + NH_4^{\oplus} \qquad (III-12)$$

Other mechanisms, however, cannot be excluded.

c. Halogenoboranes and Amines

The reaction of trichloroborane with ammonia was one of the first ever studied in the area of boron-nitrogen chemistry. However, results of such early work are not unequivocal and the reported existence of $B(NH_2)_3$ and $B_2(NH)_3$[1, 2] has not been verified (see Chapter I-D). On the other hand, as early as 1889, RIDEAL[3] obtained a white solid material on reacting trichloroborane with aniline; its analysis led to the empirical formula $BClNC_6H_5$. Fifty years later this work was repeated by JONES and KINNEY[4] who recognized the trimeric structure of the product and suggested its relation to borazine. Additional work by KINNEY and coworkers[5, 6] demonstrated that the reaction of trichloroborane with primary aromatic amines is general and leads to the formation of B-trichloro-N-triarylborazines. The study of BROWN and LAUBENGAYER[7] on the reaction of trichloroborane with ammonium chloride has already been mentioned (eq. III-2). The resultant B-trichloroborazine is one of the most convenient starting materials in borazine chemistry. Therefore this method has been developed to determine optimum conditions for synthesis. It appears that the recovery of the halogenated borazine after the reaction between trichloroborane and ammonium chloride is complete is most easily achieved by using the "dry" method of BRENNAN, DAHL and SCHAEFFER[8]. For this preparation, a Pyrex tube is filled with ammonium chloride mixed with glass beads and heated to temperatures in excess of 200° as trichloroborane is passed through; other improved procedures of comparable efficiency have been reported. For instance, EMELEUS and VIDELA[9] reported the use of metallic catalysts to improve yields of the

[1] MARTIUS, C. A.: Liebigs Ann. Chem. 109, 79 (1859).
[2] JOANNIS, A.: C. R. hebd. Séances Acad. Sci. 135, 1106 (1902).
[3] RIDEAL, S.: Ber. dtsch. chem. Ges. 22, 992 (1889).
[4] JONES, R. G., and C. R. KINNEY: J. Amer. chem. Soc. 61, 1378 (1939).
[5] KINNEY, C. R., and M. T. KOLBEZEN: ibid. 64, 1584 (1942).
[6] KINNEY, C. R., and C. L. MAHONEY: J. org. Chemistry 8, 526 (1943).
[7] BROWN, C. A., and A. W. LAUBENGAYER: J. Amer. chem. Soc. 77, 3699 (1955).
[8] BRENNAN, G. L., G. H. DAHL and R. SCHAEFFER: ibid. 82, 6248 (1960).
[9] EMELEUS, H. J., and G. J. VIDELA: J. chem. Soc. (London) 1959, 1306.

desired B-trichloroborazine and a hot-tube synthesis was described by HOHNSTEDT and LEIFIELD[1]. Today, B-trichloroborazine is commercially available.

N-Alkylated B-trichloroborazines are also readily prepared using the basic techniques of the BROWN-LAUBENGAYER procedure and utilizing prim. alkylammonium halides[2, 3]. Similarly, the use of hydrazine has been reported[4]; this procedure, however, does not yield a N-aminoborazine; instead, hydrazine decomposes during the reaction to act as a source of ammonia. Evidence was obtained for the formation of a B-trichloro-N-trihydroxyborazine upon reacting trichloroborane with hydroxylamine hydrochloride[5] but no pure material was isolated.

Tribromoborane has been used in analogous reactions to yield B-tribromoborazines. However, it has not been possible to use either trifluoroborane or triiodoborane in this reaction, although, due to the bond strength of the B—F linkage, WIBERG and HORELD[6] obtained B-trifluoro-N-trimethylborazine on condensing methylamine and dimethylfluoroborane with the evolution of methane.

The use of higher monoalkylamines in the BROWN-LAUBENGAYER synthesis requires the presence of stoichiometric amounts of a tertiary amine to facilitate the elimination of hydrogen chloride[7].

$$3\,BCl_3 + 6\,NR'_3 + 3\,RNH_2 \rightarrow (-BCl-NR-)_3 + 6[R'_3NH]Cl \quad (III\text{-}13)$$

Organodihalogenoboranes are also used in the basic preparative procedure to obtain substituted borazines through condensation of boranes with amines. For example, organodihalogenoboranes have been reacted with ammonia or primary amines to yield the corresponding B-triorganoborazines[8-13]. In many cases, the intermediate (bisamino)-organoborane can be isolated (eq. III-14).

$$RBX_2 + 4\,H_2NR' \rightarrow RB(NHR')_2 + 2[RNH_3]X \quad (III\text{-}14)$$

$$3\,RB(NHR')_2 \rightarrow (-BR-NR'-)_3 + 3\,R'NH_2 \quad (III\text{-}15)$$

[1] HOHNSTEDT, L. F., and R. F. LEIFIELD:Abstract of Papers, 133rd National Meeting of the American Chemical Society, Boston, 1959, p. 35-M.

[2] RYSCHKEWITSCH, G. E., J. J. HARRIS and H. H. SISLER: J. Amer. chem. Soc. **80**, 4515 (1958).

[3] HOHNSTEDT, L. F., and D. T. HAWORTH: ibid. **82**, 89 (1960).

[4] EMELEUS, H. J., and G. J. VIDELA: J. Chem. Soc. (London) **1959**, 1306.

[5] NIEDENZU, K., D. H. HARRELSON and J. W. DAWSON: Chem. Ber. **94**, 671 (1961).

[6] WIBERG, E., and G. HORELD: Z. Naturforsch. **6b**, 338 (1951).

[7] TURNER, H. S., and R. J. WARNE: Chem. and Ind. **1958**, 526.

[8] MIKHAILOV, B. M., and P. M. ARONOVICH: Bull. Acad. Sci. USSR., Div. Chem. Sci. **1957**, 1123.

[9] MIKHAILOV, B. M., and T. K. KOZMINSKAJA: ibid. **1958**, 597.

[10] MIKHAILOV, B. M., A. N. BLOKHINA and T. V. KOSTROMA: J. Gen. Chem. USSR. **29**, 1483 (1959).

[11] BURCH, J. E., W. GERRARD and E. F. MOONEY: J. chem. Soc. (London) **1962**, 2200.

[12] GERRARD, W., M. HOWARTH, E. F. MOONEY and D. E. PRATT: ibid. **1963**, 1582.

[13] BRAUN, J.: C. R. hebd. Séances Acad. Sci. **256**, 2422 (1963).

Closely related to this method is the preparation of borazines by the reaction of (dihalogeno)organoboranes, RBX_2, with bis(trimethylsilyl)-amine[1, 2] (eq. III-16).

$$3\,RBX_2 + 3[(CH_3)_3Si]_2NH \rightarrow (—BR—NH—)_3 + 6(CH_3)_3SiX \quad (III-16)$$

Isolation of intermediate reaction products demonstrates the mechanism of a step-wise reaction.

In a similar manner, MIKHAILOV and KOSTROMA[3] obtained B-tri-organoborazines through the reaction of alkoxy(aryl)chloroboranes with ammonia (eq. III-17).

$$3 \begin{matrix} RO \\ \diagdown \\ R' \diagup \end{matrix} B—Cl + 6NH_3 \rightarrow (—BR'—NH—)_3 + 3NH_4Cl + 3ROH \quad (III-17)$$

d. Condensation of Aminoboranes

The thermal decomposition of a variety of aminoboranes has been investigated and has been found to be a convenient method for the preparation of substituted borazines. AUBREY and LAPPERT[4] reported the formation of borazines through the pyrolysis of tris(alkylamino)-boranes and bis(alkylamino)alkoxyboranes. The pyrolysis of amino-mercaptylboranes and bis(alkylamino)phenylboranes[5] to yield substituted borazines has also been reported. An interesting borazine derivative has been obtained through the trimerization of 2-substituted 1,3,2-benzodiazaborolidines as illustrated in eq. III-18[6].

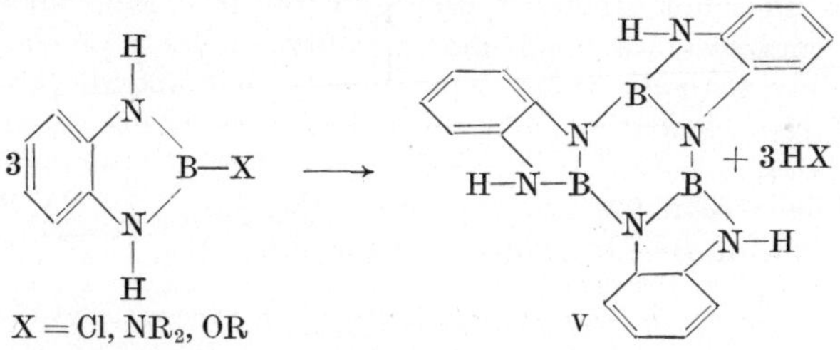

$$\text{(III-18)}$$

$$X = Cl,\ NR_2,\ OR \qquad\qquad V$$

The same compound (V) was originally obtained by the direct inter-action of trichloroborane with o-phenylenediamine at higher tempera-tures[7, 8] and was also synthesized later by reacting trialkoxyboranes, $B(OR)_3$, with o-phenylenediamine[9], the latter reaction being the first successful synthesis of a borazine derivative directly from a boron

[1] NÖTH, H.: Z. Naturforsch. **16b**, 618 (1961).
[2] JENNE, H., and K. NIEDENZU: Inorg. Chem. **3**, 68 (1964).
[3] MIKHAILOV, B. M., and T. V. KOSTROMA: Bull. Acad. Sci. USSR. **1957**, 1125.
[4] AUBREY. D. W., and M. F. LAPPERT: J. chem. Soc. (London) **1959**, 2927.
[5] BURCH, J. E., W. GERRARD and E. F. MOONEY: ibid. **1962**, 2200.
[6] BEYER, H., K. NIEDENZU and J. W. DAWSON: J. org. Chemistry **27**, 4701 (1962).
[7] SCHUPP, L. J., and C. A. BROWN: Abstract of Papers, 128th National Meeting of the American Chemical Society, Minneapolis, 1955, p. 48-R.
[8] RUDNER, B., and J. J. HARRIS: Abstract of Papers, 138th National Meeting of the American Chemical Society, New York, 1960. p. 61-P.
[9] BROTHERTON, R. J., and H. STEINBERG: J. org. Chemistry **26**, 4632 (1961).

ester. Nitrogen-substituted bisaminoboranes react with anhydrous ammonia to yield borazines, even at low temperatures[1]. This behavior

$$3\,RB(NHR')_2 + 3\,NH_3 \rightarrow (\text{---}BR\text{---}NH\text{---})_3 + 6\,R'NH_2 \qquad (\text{III-19})$$

is consistent with the fact that bisaminoboranes with free NH_2 groups are unstable and readily condense to borazines with the elimination of ammonia.

2. Replacement Reactions of the Borazine Ring

a. Substitution of B-bonded Hydrogen

SCHLESINGER, RITTER and BURG[2] have investigated the reaction of borazine and N-trimethylborazine with trimethylborane. Besides methyldiboranes, a mixture of B-methylated borazine derivatives was obtained, indicating the exchange of the boron-attached hydrogen of the borazines with the methyl groups of the trimethylborane. More practical methods for alkylation and arylation of the B—H group in borazines have since been developed. Both GRIGNARD and organolithium reagents give satisfactory yields of the desired products[3] according to the equation:

$$(\text{---}BH\text{---}NR\text{---})_3 + 3\,LiR' \rightarrow (\text{---}BR'\text{---}NR\text{---})_3 + 3\,LiH \qquad (\text{III-20})$$

Replacement reactions of the B-attached hydrogen of N-substituted borazines have not otherwise been explored in detail. Since the B-trihalogenoborazines are much more readily available and are therefore used mainly for subsequent replacement reactions, the lack of effort to effect replacement of B-attached hydrogen is understandable.

The bromination of borazine to yield 2,4-dibromoborazine has been cited earlier (see p. 92) and represents the sole example of the direct halogenation of B—H groups in borazines.

b. Substitution of B-bonded Halogen

α) *Replacement of halogen by hydrogen.* The reduction of B-trichloroborazine to the parent compound has already been discussed. Analogous procedures have been applied to N-substituted borazines. Difficulties met in isolating the products of reduction of B-trichloroborazine with $LiAlH_4$[4] are not found on reducing the N-substituted compounds[3]. However, the most convenient route seems to involve reduction with sodium hydroborate as developed by HOHNSTEDT and HAWORTH[5].

[1] NIEDENZU, K., and coworkers: unpublished results.
[2] SCHLESINGER, H. I., D. M. RITTER and A. B. BURG: J. Amer. chem. Soc. 60, 1296, 2297 (1938).
[3] SMALLEY, J. H., and S. F. STAFIEJ: ibid. 81, 582 (1959).
[4] SCHAEFFER, R., M. STEINDLER, L. F. HOHNSTEDT, H. S. SMITH JR., L. B. EDDY and H. I. SCHLESINGER: ibid. 76, 3303 (1954).
[5] HOHNSTEDT, L. F., and D. T. HAWORTH: ibid. 82, 89 (1960).

β) *Transhalogenation.* Transhalogenation of B-trichloroborazines has recently attracted attention. Since B-trifluoroborazines are not accessible through the BROWN-LAUBENGAYER method[1], transhalogenation reactions involving several fluorinating agents were investigated. NIEDENZU[2] reported the fluorination of B-trichloroborazine with a variety of metal fluorides (eq. III-21). The best results were obtained

$$(—BCl—NH—)_3 + 3\,MeF \rightarrow (—BF—NH—)_3 + 3\,MeCl \qquad (III\text{-}21)$$

utilizing titanium tetrafluoride; the method was extended to a variety of N-substituted borazines[3] and metal iodides and azides were used in an analogous reaction as described by eq. III-21[4,5]. Transhalogenation with SbF_3 has also been employed successfully to prepare B-trifluoroborazine[6] and mixtures of SbF_3 with antimony chlorides were used to synthesize 2,4-difluoro-6-chloroborazine, VI, and 2-fluoro-4,6-dichloroborazine, VII[7].

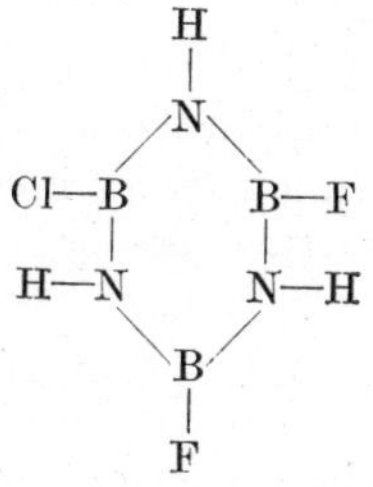

VI 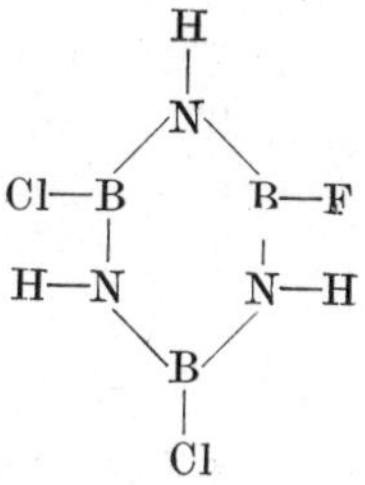   VII

γ) *B-Alkylation and Arylation.* The substitution of boron-attached halogen by organic groups was simultaneously developed in several laboratories. It is most easily achieved by the GRIGNARD method[8-11]

$$(—BCl—NR—)_3 + 3\,R'MgX \rightarrow (—BR'—NR—)_3 + 3\,MgXCl \qquad (III\text{-}22)$$

but has also been effected through the use of organo-lithium compounds[12].

[1] BROWN, C. A., and A. W. LAUBENGAYER: J. Amer. chem. Soc. **77**, 3699 (1955).
[2] NIEDENZU, K.: Inorg. Chem. **1**, 943 (1962).
[3] NIEDENZU, K., H. BEYER and H. JENNE: Chem. Ber. **96**, 2649 (1963).
[4] MUSZKAT, K. A., L. HILL and B. KIRSON: Israel J. Chem. **1**, 27 (1963).
[5] MUSZKAT, K. A., and B. KIRSON: ibid. **1**, 150 (1963).
[6] LAUBENGAYER, A. W., K. WATTERSON, D. R. BIDINOSTI and R. F. PORTER: Inorg. Chem. **2**, 519 (1963).
[7] BEYER, H., J. B. HYNES, H. JENNE and K. NIEDENZU: Advances in Chemistry **42**, 266 (1964).
[8] RYSCHKEWITSCH, G. E., J. J. HARRIS and H. H. SISLER: J. Amer. chem. Soc. **80**, 4515 (1958).
[9] BECHER, H. J., and S. FRICK: Z. anorg. allg. Chem. **295**, 83 (1958).
[10] GROSZOS, A., and S. F. STAFIEJ: J. Amer. chem. Soc. **80**, 1357 (1958).
[11] PELLON, J., W. G. DEICHERT and W. M. THOMAS: J. Polymer. Sci. **55**, 153 (1961).
[12] SMALLEY, J. H., and S. F. STAFIEJ: J. Amer. chem. Soc. **81**, 582 (1959).

By adjusting the reagent proportions, it has been found possible to

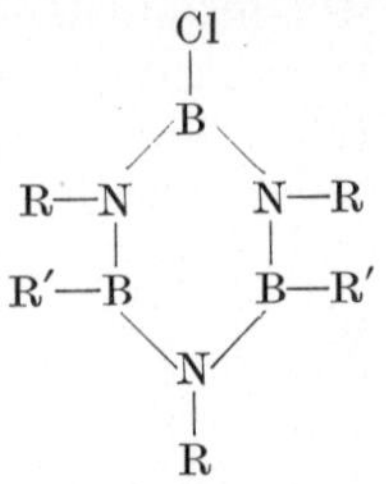

obtain mixed B-organo-chloro derivatives, VIII and IX,[1,2] but no derivatives containing perfluoroalkyl groups could be prepared by either method[3].

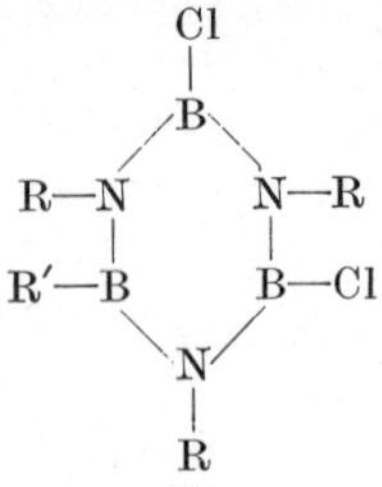

A brief report by NIEDENZU and DAWSON indicates that B-trichloroborazine can participate in the FRIEDEL-CRAFTS reaction with benzene and aluminium chloride in boiling chlorobenzene to yield B-triphenylborazine[4] but this reaction has not been explored in detail.

δ) *Aminolysis.* The aminolysis of B-trichloroborazines was first reported in a patent[5] and has since been exhaustively investigated[6-9]; a variety of primary and secondary amines has successfully been employed in this reaction.

$$(-BCl-NH-)_3 + 6 R_2NH \rightarrow (-BNR_2-NH-)_3 + 3 [R_2NH_2]Cl \ (III\text{-}23)$$

On ammonolysis of B-trichloroborazine, however, only unidentified, apparently polymeric materials were obtained[6]. B-Triaminoborazine, $(-BNH_2-NH-)_3$, was later prepared by a transamination reaction[7,10]. In an analogous fashion B-trishydrazinoborazine has been obtained[10].

$$(-BNR_2-NH-)_3 + 3 H_2NZ \rightarrow (-BNHZ-NH-)_3 + 3 R_2NH \ (III\text{-}24)$$
$$Z = H, NH_2$$

Transamination can likewise be effected using substituted hydrazines and primary and secondary amines[7,11,12]. In this connection, it is of interest

[1] RYSCHKEWITSCH, G. E., J. J. HARRIS and H. SISLER: J. Amer. chem. Soc. **80,** 4515 (1958).

[2] MELLER, A.: Mh. Chem. **94,** 183 (1963).

[3] LAGOWSKI, J. J., and P. G. THOMPSON: Proc. chem. Soc. (London) **1959,** 301.

[4] NIEDENZU, K., and J. W. DAWSON: Angew. Chem. **71,** 651 (1959).

[5] GOULD, J. R.: U.S. Patent 2,754,177 (1956).

[6] NIEDENZU, K., and J. W. DAWSON: J. Amer. chem. Soc. **81,** 3561 (1959).

[7] GERRARD, W., H. R. HUDSON and E. F. MOONEY: J. chem. Soc. (London) **1962,** 113.

[8] TOENISKOETTER, R. H., and F. R. HALL: Inorg. Chem. **2,** 29 (1963).

[9] GUTMANN, V., A. MELLER and R. SCHLEGEL: Mh. Chem. **94,** 1071 (1963).

[10] NIEDENZU, K., and J. W. DAWSON: Angew. Chem. **73,** 433 (1961).

[11] NIEDENZU, K., D. H. HARRELSON and J. W. DAWSON: Chem. Ber. **94,** 671 (1961).

[12] NIEDENZU, K., W. GEORGE and J. W. DAWSON: Abstract of Papers, 141st National Meeting of the American Chemical Society, Chicago 1961, p. 15-N.

to note that primary and secondary amines will react with B-trialkyl-borazines and will cleave the borazine ring[1].

$$(—BR—NR'—)_3 + 6R''NH_2 \rightarrow 3RB(NHR'')_2 + 3R'NH_2 \quad \text{(III-25)}$$

The reaction of B-aminoborazines which produces cleavage of the exocyclic B—N linkage with trichloroborane, phosphorus oxychloride, hydrogen chloride, phosphorus trichloride and alcohols has been investigated[2, 3, 4]. When stoichiometric amounts of the reagents are used, the borazine ring remains intact. This method has been found most useful for the synthesis of perfluoroalkoxyborazines[5, 6].

Two different mechanisms have been observed when B-trisamino-borazines react with organic isocyanates. On treating B-tris(diethylamino)borazine with isocyanates in inert solvents, migration of the exocyclic amino group occurs with the formation of a B-trisureido-borazine.

$$[—BN(C_2H_5)_2—NH—]_3 + 3RNCO \rightarrow \{—B[NR—CO—N(C_2H_5)_2]—NH—\}_3$$
$$\text{(III-26)}$$

On the other hand, when B-tris(diethylamino)-N-triethylborazine is reacted with an excess of phenyl isocyanate, cleavage of the borazine ring and subsequent formation of a new ring system ensues[7, 8].

$$[—BN(C_2H_5)_2—NC_2H_5]_3 + 6RNCO \longrightarrow 3$$

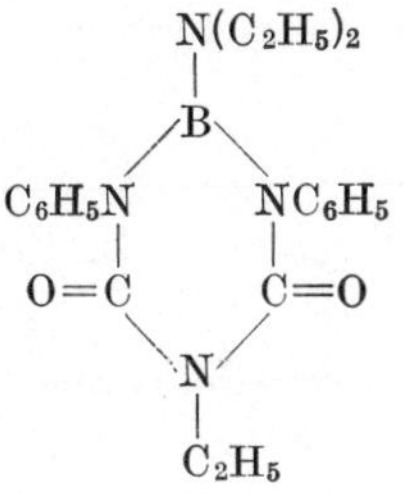

$$R = C_6H_5 \qquad \qquad \text{(III-27)}$$

ε) *Metatheses with metal salts.* Metathetical reactions between B-trichloroborazines and a variety of metal salts have been explored; the transhalogenation reaction has been described above. However, it has been found possible to replace the boron-attached halogen by alkoxy,

[1] ENGLISH, W. D., A. L. McCLOSKEY and H. STEINBERG: J. Amer. chem. Soc. **83**, 2122 (1961).

[2] NIEDENZU, K., D. H. HARRELSON and J. W. DAWSON: Chem. Ber. **94**, 671 (1961).

[3] GERRARD, W., H. R. HUDSON and E. F. MOONEY: J. chem. Soc. (London) **1962**, 113.

[4] TOENISKOETTER, R. H., and F. R. HALL: Inorg. Chem. **2**, 29 (1963).

[5] NIEDENZU, K., W. GEORGE and J. W. DAWSON: Abstract of Papers, 141st National Meeting of the American Chemical Society, Chicago 1961, p. 15-N.

[6] NIEDENZU, K., J. W. DAWSON and W. GEORGE: Chem. and Ind. **1963**, 255.

[7] CRAGG, R. H., and M. F. LAPPERT: J. chem. Soc. (London) **1964**, 2108.

[8] BEYER, H., J. W. DAWSON, H. JENNE and K. NIEDENZU: ibid. **1964**, 2115.

aryloxy[1], CN, CNO, CNS[2-4] and N_3 [5] groups by reacting B-trichloro-borazines with the sodium, potassium, lead or silver salts of the appropriate anion. By properly adjusting the stoichiometry of the reactants, it was found possible to obtain unsymmetrically substituted products[3].

ζ) *Some other reactions of B-halogenated borazines.* B-Trichloro-borazines form 1 : 3 addition complexes with phosphoric esters. On heating the addition compounds in organic solvents they decompose yielding phosphorylated borazines[6].

$$(-BCl-NR-)_3 + 3 OP(OR')_3 \rightarrow [-BOP(O)(OR')_2-NR-]_3 + 3 R'Cl \quad (III-28)$$

With diazomethane, B-trichloroborazine forms chloromethyl derivatives while preserving the six-membered ring[7].

$$(-BCl-NH-)_3 + 3 CH_2N_2 \rightarrow (-BCH_2Cl-NH-)_3 + 3 N_2 \quad (III-29)$$

Closely related to this reaction is that of B-trichloroborazines with organic oxides yielding β-chlorinated alkoxyborazines[8].

$$(-BCl-NH-)_3 + 3 (CH_2)_2O \rightarrow (-BOCH_2CH_2Cl-NH-)_3 \quad (III-30)$$

Borazines containing an exocyclic boron-silicon bond $(-BSiR_3-NR'-)_3$, or boron-oxygen-silicon linkage, $(-BOSiR_3-NR'-)_3$, were obtained by reacting B-trichloroborazines with the proper organometallic derivatives[9-11].

3. Properties of Borazines

The four B-halogenated borazines, $(-BX-NH-)_3$ with X = F, Cl, Br, and I, have been described (see Table III-5). They are not available by direct halogenation of borazine. As a matter of fact, the halogenation of borazine has been investigated only twice. WIBERG and BOLZ[12] studied the bromination of borazine (see Chapter III-B) and obtained 2,4-dibromoborazine. The direct iodination of borazine[13] resulted in

[1] BRADLEY, M. J., G. E. RYSCHKEWITSCH and H. H. SISLER: J. Amer. chem. Soc. **81**, 2635 (1959).
[2] BRENNAN, G. L., G. H. DAHL and R. SCHAEFFER: ibid. **82**, 6248 (1960).
[3] LAPPERT, M. F., and H. PYSZORA: J. chem. Soc. (London) **1963**, 1744.
[4] MELLER, A.: Personal communication.
[5] MUSZKAT, K. A., L. HILL and B. KIRSON: Israel J. Chem. **1**, 27 (1963).
[6] NIEDENZU, K., and J. W. DAWSON: Angew. Chem. **72**, 920 (1960).
[7] TURNER, H. S.: Chem. and Ind. **1958**, 1405.
[8] NIEDENZU, K., J. W. DAWSON and W. GEORGE: ibid. **1963**, 255.
[9] SEYFERTH, D., and H. P. KÖGLER: J. inorg. nucl. Chem. **15**, 99 (1960).
[10] COWLEY, A. H., H. H. SISLER and G. E. RYSCHKEWITSCH: J. Amer. chem. Soc. **82**, 501 (1960).
[11] MELLER, A.: Mh. Chem. **94**, 183 (1963).
[12] WIBERG, E. and A. BOLZ: Ber. dtsch. chem. Ges. **73**, 209 (1940).
[13] MUSZKAT, K. A. and B. KIRSON: Israel J. Chem. **1**, 150 (1963).

the formation of a non-volatile unidentified product of a probable polymeric nature. With the exception of the iodo compound, B-trihalogenoborazines are thermally quite stable under anhydrous conditions.

Table III-5. *B-Halogenated Borazines*, $(-BX-NH-)_3$

Borazine	References	Borazine	References
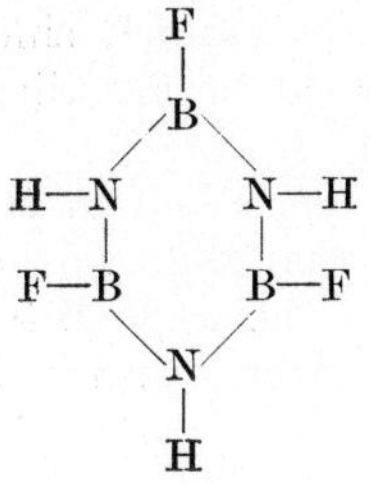 F | B H—N⟋ ⟍N—H F—B B—F ⟍N⟋ | H m.p. 122°	1, 2	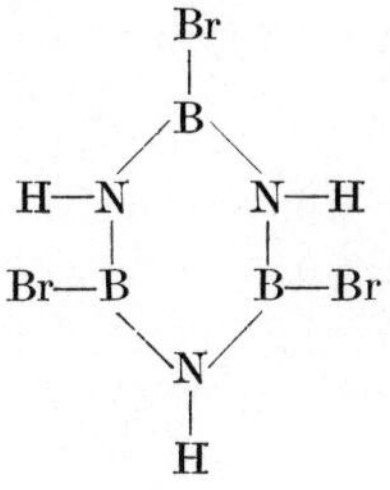 Br | B H—N⟋ ⟍N—H Br—B B—Br ⟍N⟋ | H m.p. 131°	5, 6, 7
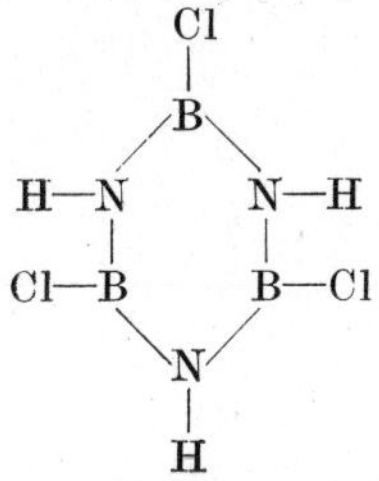 Cl | B H—N⟋ ⟍N—H Cl—B B—Cl ⟍N⟋ | H m.p. 84°	3, 4, 5	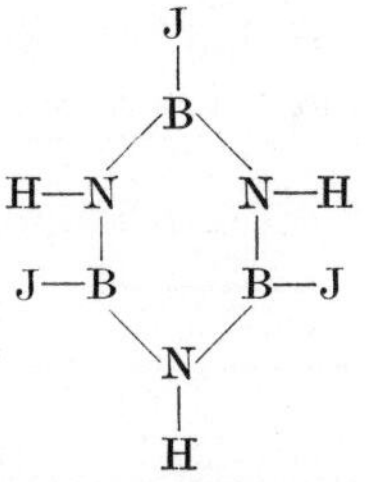 J | B H—N⟋ ⟍N—H J—B B—J ⟍N⟋ | H m.p. 134°	7

B-Triiodoborazine, however, decomposes at room temperature within a few days[8]. A wide variety of N-substituted B-trihalogenoborazines are known. The B-chloro and B-bromo derivatives are readily obtained by the BROWN-LAUBENGAYER synthesis[4], but, with the exception of

[1] NIEDENZU, K.: Inorg. Chem. **1**, 943 (1962).
[2] LAUBENGAYER, A.W., K. WATTERSON, D. R. BIDINOSTI and R. F. PORTER: ibid. **2**, 519 (1963).
[3] STOCK, A., E. WIBERG and H. MARTINI: Ber. dtsch. chem. Ges. **63**, 2927 (1930).
[4] BROWN, C. A., and A. W. LAUBENGAYER: J. Amer. chem. Soc. **77**, 3699 (1955).
[5] EMELEUS, H. J., and G. J. VIDELA: J. chem. Soc. (London) **1959**, 1306.
[6] SCHAEFFER, R., M. STEINDLER, L. F. HOHNSTEDT, H. S. SMITH Jr., L. B. EDDY and H. I. SCHLESINGER: J. Amer. chem. Soc. **76**, 3303 (1954).
[7] NÖTH, H.: Z. Naturforsch. **16b**, 618 (1961).
[8] NÖTH, H.: Personal communication.

B-trifluoro-N-trimethylborazine[1], B-fluorinated derivatives generally have been obtained only by means of halogen exchange reactions[2-5]. However, HARRIS and RUDNER[6] have developed a new approach. They found that alkylamine-trifluoroboranes, $RH_2N \cdot BF_3$, can indeed be dehydrohalogenated to yield $(-BF-NR-)_3$ by reaction with a tri-substituted amine-trifluoroborane, $R_3'N \cdot BF_3$, if the substituents R' are sterically encumbered. Under such circumstances, reaction occurs in analogy to eq. (III-13), apparently being promoted by the tendency of the amine-borane $R_3'N \cdot BF_3$ to convert to an ammonium fluoroborate, $[R_3'NH]BF_4$. N-Substituted borazines with hydrogen attached to the boron are conveniently iodinated by elemental iodine[7], since halogen exchange reactions produced only low yields of the described materials.

Reactions of B-halogenated borazines have been described in the previous chapter. In addition, it is worth mentioning that B-alkylation cannot be achieved with aluminum trialkyl. For example, when B-tri-chloroborazine was reacted with triethylaluminium, an exothermic reaction illustrated in eq. III-30 was observed[8].

$$(-BCl-NH-)_3 + 3\,AlR_3 \rightarrow 3\,BR_3 + (-AlCl-NH-)_x \qquad (III-31)$$

Borazines with exocyclic B—N—P linkages have been prepared from B-trichloroborazines and N-metallated phosphorus amides or from B-aminoborazines and phosphorus halides[9].

Table III-6. *B-Trifluoroborazines, $(-BF-NR-)_3$*

Borazine	m.p. °C	b.p.(mm.) °C	References
$(-BF-NCH_3-)_3$	90.5		1, 3, 10
$(-BF-NC_2H_5-)_3$		26 (3)	3
$(-BF-N-n-C_3H_7-)_3$		59 (3)	3
$(-BF-N-n-C_4H_9-)$		89 (3)	3
$(-BF-NCH_2C_6H_5-)_3$	107	215 (0.4)	4
$(-BF-NC_6H_5-)_3$	154	200 (0.5)	4
$(-BF-NSiH_3-)_3$		24 (2.5)	10

[1] WIBERG, E., and G. HORELD: Z. Naturforsch. **6b**, 338 (1951).
[2] NIEDENZU, K.: Inorg. Chem. **1**, 943 (1962).
[3] NIEDENZU, K., H. BEYER and H. JENNE: Chem. Ber. **96**, 2649 (1963).
[4] MUSZKAT, K. A., L. HILL and B. KIRSON: Israel J. Chem. **1**, 27 (1963).
[5] LAUBENGAYER, A. W., K. WATTERSON, D. R. BIDINOSTI and R. F. PORTER: Inorg. Chem. **2**, 519 (1963).
[6] HARRIS, J. J., and B. RUDNER: Abstract of Papers, 147th National Meeting of the American Chemical Society, Philadelphia 1964, p. 25-L.
[7] MUSZKAT, K. A., and B. KIRSON: Israel J. Chem. **1**, 150 (1963).
[8] WOODS, W. G., and A. L. McCLOSKEY: Inorg. Chem. **2**, 861 (1963).
[9] GUTMANN, V., A. MELLER and R. SCHLEGEL: Mh. Chem. **94**, 733 (1963).
[10] SUJISHI, S., and S. WITZ: J. Amer. chem. Soc. **79**, 2447 (1957).

Table III-7. *B-Trichloroborazines*, $(-BCl-NR-)_3$

Borazines	m.p. °C	b.p. (mm.) °C	References
$(-BCl-NCH_3-)_3$	162—164		3, 7, 8, 9
$(-BCl-NC_2H_5-)_3$	57—59		3, 9
$(-BCl-N-n-C_3H_7-)_3$	79—80	110 (0.01)	6
$(-BCl-N-n-C_4H_9-)_3$	30	115—120 (0.5)	9
$(-BCl-N-s-C_4H_9-)_3$		102 (0.005)	6
$(-BCl-N-cyclo-C_6H_{11}-)_3$	217—219		3
$(-BCl-NC_6H_5-)$	273—275		3, 4
$(-BCl-N-p-CH_3C_6H_4-)_3$	308—309		1, 2, 5
$(-BCl-N-p-CH_3OC_6H_4-)_3$	233—238		1, 2, 3
$(-BCl-N-o-CH_3C_6H_4-)_3$	198—202		1
$(-BCl-N-m-CH_3C_6H_4-)_3$	269—271		1
$(-BCl-N-p-BrC_6H_4-)_3$			1, 2

Table III-8. *B-Tribromoborazines*, $(-BBr-NR-)_3$

Borazines	m.p. °C	References
$(-BBr-NC_2H_5-)_3$	78—82	3
$(-BBr-NC_6H_5-)_3$	292—293	10, 11

Table III-9. *B-Triiodoborazines*, $(-BI-NR-)_3$

Borazines	m.p. °C	References
$(-BI-N-n-C_4H_9-)_3$	148—150	12
$(-BI-NC_6H_5-)_3$	118	13

A number of B-tripseudohalogenoborazines have been prepared. They have been shown to be useful intermediates for synthetic studies[6] and some N-hydrogenated derivatives are listed in Table III-10.

Table III-10. *B-Tripseudohalogenoborazines*, $(-BX-NH-)_3$

Borazines	m.p. °C	References
$(-BCN-NH-)_3$	200 (d.)	14
$(-BNCO-NH-)_3$	166	6
$(-BNCS-NH-)_3$	154 (d.)	6, 14
$(-BN_3-NH-)_3$		13

[1] GERRARD, W., E. F. MOONEY and D. E. PRATT: J. appl. Chem. **13**, 127 (1963).
[2] GERRARD, W., and E. F. MOONEY: J. chem. Soc. (London) **1960**, 4028.
[3] HOHNSTEDT, L. F., and D. T. HAWORTH: J. Amer. chem. Soc. **82**, 89 (1960).
[4] JONES, R. G., and C. R. KINNEY: ibid. **61**, 1378 (1939).
[5] KINNEY, C. R., and M. T. KOLBEZEN: ibid. **64**, 1584 (1942).
[6] LAPPERT, M. F., and H. PYSZORA: J. chem. Soc. (London) **1963**, 1744.
[7] NIEDENZU, K., and J. W. DAWSON: J. Amer. chem. Soc. **81**, 3561 (1959).
[8] RYSCHKEWITSCH, G. E., J. J. HARRIS and H. H. SISLER: ibid. **80**, 4515 (1958).
[9] TURNER, H. S., and R. J. WARNE: Chem. and Ind. **1958**, 526.
[10] MIKHAILOV, B. M., A. N. BLOKHINA and T. V. KOSTROMA: J. Gen. Chem. USSR. **29**, 1483 (1959).
[11] MIKHAILOV, B. M., and T. V. KOSTROMA: ibid. **29**, 1477 (1959).
[12] MUSZKAT, K. A., L. HILL and B. KIRSON: Israel J. Chem. **1**, 27 (1963).
[13] MUSZKAT, K. A., and B. KIRSON: ibid. **1**, 150 (1963).
[14] BRENNAN, G. L., G. H. DAHL and R. SCHAEFFER: J. Amer. chem. Soc. **82**, 2648 (1960).

The thermal degradation of a series of organo-substituted borazines was studied in the temperature range between 370—525° in order to determine the effect of substituents on the stability of the borazine nucleus and the mode of pyrolytic decomposition[1]. The results are illustrated in Table III-11.

Table III-11. *Thermal Stability of Substituted Borazines, $(-BR-NR'-)_3$ at 450°*

Substituents	Increase in thermal stability				
	1.	2.	3.	4.	5.
R	C_6H_5	C_6H_5	CH_3	C_6H_5	CH_3
R′	CH_3	C_6H_5	C_6H_5	H	CH_3

Analysis of the products indicates that boron-carbon homolysis is the primary step in most cases, although decomposition of B-triphenyl-N-triphenylborazine may proceed through an initial boron-nitrogen cleavage. A considerable amount of benzene was obtained in those cases where a B-phenyl linkage was present, whereas, on decomposition of B-trimethyl-N-triphenylborazine, no aromatic product was observed. It can be inferred that the diradical $C_6H_5\dot{N}\cdot$ is first formed and this entity dimerizes to yield azobenzene. Elimination of nitrogen should result in the formation of biphenyl and indeed this material was isolated from the reaction mixture.

The thermal decomposition of B-triamino-N-triphenylborazine in vacuo has been reported to yield boron nitride and aniline[2]. This behavior indicates a greater thermal stability of the exocyclic B—N bond in B-aminoborazines. In contrast, on hydrolytic attack, the primary step of the reaction appears to be the rupture of the exocyclic bond, although the ring bonds do not appear to be much more stable[3].

A detailed study of exchange reactions between differently substituted borazines indicates the existence of an apparent equilibrium between B-trimethyl-N-trimethylborazine, B-trichloro-N-trimethylborazine and the 2,4-dichloro-6-methyl and 2-chloro-4,6-dimethyl derivatives. Similarly, B-trimethyl-N-trimethylborazine and N-trimethylborazine can exchange boron substituents slowly at 175°, but the reaction does not appear to be of preparative value[4]. Boron and nitrogen substituents exchange was also reported by WAGNER and BRADFORD[5]; it is quite possible that the reactions proceed via a ring cleavage mechanism and random recombination of monomer fragments.

Some reactions of B-vinylated borazines have been investigated for the purpose of synthesizing polymeric materials. B-Trivinyl-N-triphenyl-

[1] NEWSOM, H. C., W. D. ENGLISH, A. L. McCLOSKEY and W. G. WOODS: J. Amer. chem. Soc. **83,** 4134 (1961).
[2] TOENISKOETTER, R. H., and F. R. HALL: Inorg. Chem. **2,** 29 (1963).
[3] NIEDENZU, K., and J. W. DAWSON: J. Amer. chem. Soc. **81,** 3561 (1959).
[4] NEWSOM, H. C., W. G. WOODS and A. L. McCLOSKEY: Inorg. Chem. **2,** 36 (1963).
[5] WAGNER, R. I., and J. L. BRADFORD: ibid. **1,** 99 (1962).

borazine, **XI**, has not yet been homopolymerized. On radical-initiated copolymerisation with other vinyl monomers, the B-vinylated borazine is competitive with styrene in reactivity[1]. SEYFERTH and TAKAMIZAWA[2]

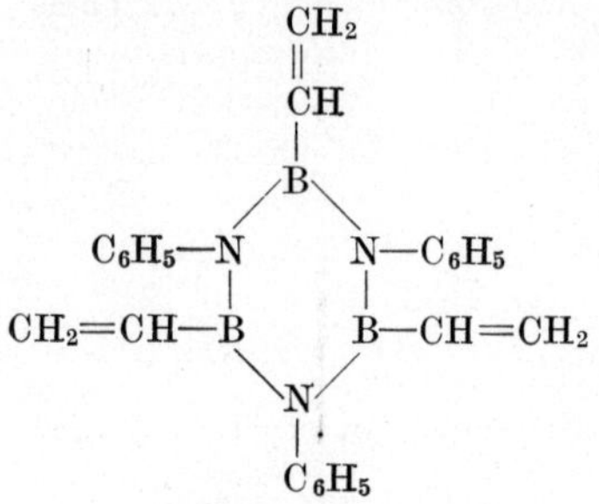

$$CH_2$$
$$\|$$
$$CH$$
$$|$$
$$B$$

$$C_6H_5-N \qquad N-C_6H_5$$
$$CH_2=CH-B \qquad B-CH=CH_2$$
$$N$$
$$|$$
$$C_6H_5$$

XI

reported the addition of a number of reagents to the vinyl groups of B-vinylated borazines in the presence of radical initiators. For instance smooth addition of CBr_4 was achieved in the presence of benzoyl peroxide.

$$3\,CBr_4 + (-BCHCH_2-NC_6H_5-)_3 \rightarrow (-BCHBrCH_2CBr_3 - NC_6H_5-)_3 \quad \text{(III-32)}$$

B-Trivinylborazine[3] polymerizes on standing, even if sealed under vacuum[4]. The resultant product has not yet been investigated.

The hydrolytic stability of the borazine ring is influenced by the substituents on both the boron and the nitrogen atoms. In agreement with some theoretical considerations[5], it is apparent that the influence of the boron substituents is much less pronounced[6] than is the case with the nitrogen substituents. In general, however, borazines do not show appreciable stability towards hydrolytic attack[7].

LAUBENGAYER and coworkers[8] gathered some mass spectral data which enables one to compare the stability of the borazine ring system as it exists in B-trifluoroborazine and B-trichloroborazine with B-trifluoroboroxine, **XII**.

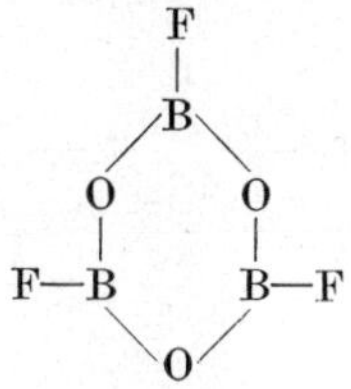

$$F$$
$$|$$
$$B$$
$$O \qquad O$$
$$|\qquad\quad|$$
$$F-B \qquad B-F$$
$$O$$

XII

[1] PELLON, J., W. G. DEICHERT and W. M. THOMAS: J. Polymer. Sci. **55**, 153 (1961).
[2] SEYFERTH, D., and M. TAKAMIZAWA: J. org. Chemistry **28**, 1142 (1963).
[3] FRITZ, P., K. NIEDENZU and J. W. DAWSON: Inorg. Chem. **3**, 626 (1964).
[4] NIEDENZU, K., and coworkers: unpublished results.
[5] DAVIES, D. W.: Trans. Faraday Soc. **56**, 1713 (1960).
[6] MELLER, A.: Mh. Chem. **94**, 183 (1963).
[7] BROTHERTON, R.J., and A.L. McCLOSKEY: Advances in Chemistry **42**, 131 (1964).
[8] LAUBENGAYER, A. W., K. WATTERSON, D. R. BIDINOSTI and R. F. PORTER: Inorg. Chem. **2**, 519 (1963).

The parent ion was predominant in all compounds. This feature is characteristic of aromatic ring systems, whereas, for halogenated derivatives of saturated hydrocarbons, the major ion usually results from the loss of one halogen atom. On the basis of the mass spectral data, it was concluded that B-trifluoroborazine is more likely to undergo ring cleavage to form higher condensed heterocyclic polymers than to split off hydrogen fluoride intermolecularly to' form diborazinyl

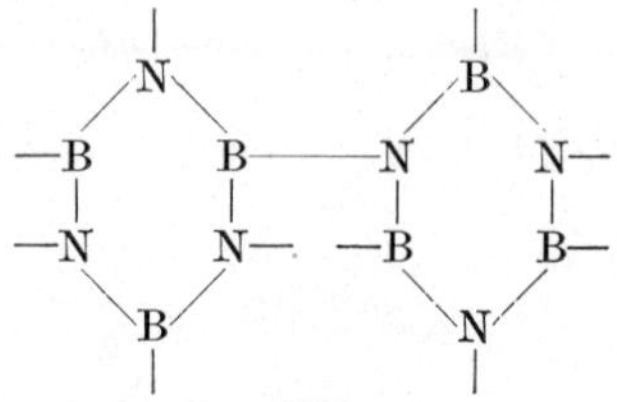

XIII

compounds, XIII. The latter, however, are more likely to be found among the products of condensation of B-trichloroborazine due to facile intermolecular elimination of HCl without ring cleavage.

At its melting point, B-tris(methylamino)-N-triphenylborazine was observed to transform irreversibly to another molecule of identical analytical composition[1]. Similarly, it appears that B-trichloro-N-triphenylborazine exists in two forms which have not been further characterized.

There is considerable doubt that the reaction of B-triorganoborazines, (—BR—NH—)$_3$, with isocyanates proceeds as originally formulated by KORSHAK and coworkers[2] through addition of the isocyanate across the N—H bond of the borazine. BOONE and WILLCOCKSON[3] demonstrated that cleavage of the borazine ring occurs and, upon rearrangement, with phenyl isocyanate the new cyclic ring system XIV is obtained.

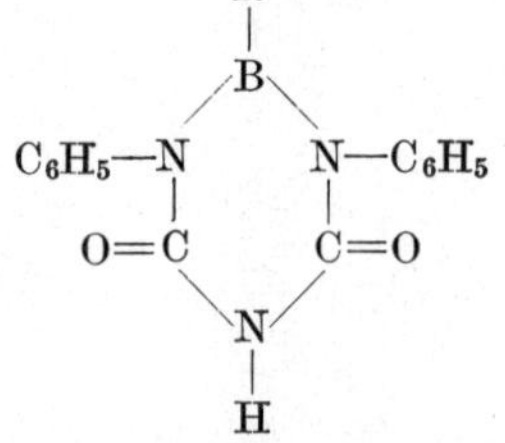

XIV

With the exception of the apparent formation of an impure B-trichloro-N-trihydroxyborazine on reacting trichloroborane with

[1] TOENISKOETTER, R. H., and F. R. HALL: Inorg. Chem. 2, 29 (1963).

[2] KORSHAK, V. V., N. I. BEKASOVA, V. A. ZAMYATINA and G. I. ARISTORKHOVA: Vysokomolekulyarnye Soedineniya 3, 525 (1961).

[3] BOONE, J. L., and G. W. WILLCOCKSON: Abstr. of Papers, 142nd National Meeting of the American Chemical Society, Atlantic City 1962, p. 6-N.

hydroxylamine[1] and the recently reported hexachloroborazine, (—BCl—NCl—)$_3$ [2], N-substituted borazines with other than alkyl or aryl substituents are unknown. In this connection, it is of extreme interest that recently N-lithio substituted borazines have been prepared[3]. Such compounds are formed by reacting borazines containing N—H groups with lithium alkyl in ether. The resultant N-lithio derivatives have so far been used only for the preparation of polynucleated borazines (see Chapter III-E) and borazines with unsymmetrical N-substitution. Interesting aspects can be foreseen from a more detailed study of these metallated borazines.

Mechanistic studies of the reactions of borazines have hardly been implemented so far. Since the borazine ring possesses both, acceptor and donor sites, addition at a borazine nucleus can proceed by various routes. In general, borazines form adducts with such materials as water, alcohols, and hydrogen halides in a 1:3 molar ratio even when one of the reactants is present in excess. Such donor molecules, as the ones cited above, are known to form adducts in other systems by ionic dissociation. It appears reasonable to assume electron donor addition at the boron site to explain their interaction with borazines. The thermal decomposition of these adducts can occur by any of three ways:

1. Hydrogen halides are eliminated with preservation of the original ring system. However, the stereochemistry of the mechanism for the addition and elimination are different. This was illustrated by LAUBEN-GAYER and coworkers[4] in a study of the reaction of borazine with deutero halides.

2. Pyrolysis of the methanol adduct of borazine was shown to provide for an exchange of the boron hydrogens but with preservation of the original boron-nitrogen ring to yield B-trimethoxyborazine[5].

3. Pyrolysis of water adducts while in the mass spectrophotometer showed that the boron-nitrogen ring is largely destroyed[6, 7].

However, in a few cases, 1:1 addition to a borazine ring has been observed. There is some evidence that such addition indicates the formation of a charge transfer complex, such as in the case of tetra-cyanoethylene (see also Chapter III-C 5), which might also be the case with the reported adducts of iodine[8] and bromine[9]. In this connection, it is of interest to note that several borazines are reported to crystallize

[1] NIEDENZU, K., D. H. HARRELSON and J. W. DAWSON: Chem. Ber. **94,** 671 (1961).

[2] PAETZOLD, P. I.: Z. anorg. allg. Chem. **326,** 47 (1963).

[3] WAGNER, R. I., and J. L. BRADFORD: Inorg. Chem. **1,** 93 (1962).

[4] LAUBENGAYER, A. W.: Personal communication.

[5] HAWORTH, D. T., and L. F. HOHNSTEDT: J. Amer. Chem. Soc. **81,** 842 (1959).

[6] WEIBRECHT, W. E., and A. W. LAUBENGAYER: Abstr. of Papers, 145th National Meeting of the American Chemical Society, New York 1963, p. 4-N.

[7] WEIBRECHT, W. E.: Ph. D. Thesis, Cornell University, 1964.

[8] MUSZKAT, K. A., and B. KIRSON: Israel J. Chem. **1,** 150 (1963).

[9] WIBERG, E., and A. BOLZ: Ber. dtsch. chem. Ges. **73,** 209 (1940).

from aromatic solvents as adducts[1-5]. However, more information
is needed before any definite conclusions can be drawn. Only the 1:1
adduct of trimethylamine and borazine has been studied in detail. The
adduct is believed to result from the formation of a donor-acceptor
complex bond between one boron ring site and the nitrogen of the
trimethylamine, XV. Pyrolysis of the adduct at 60—100° yields borazine

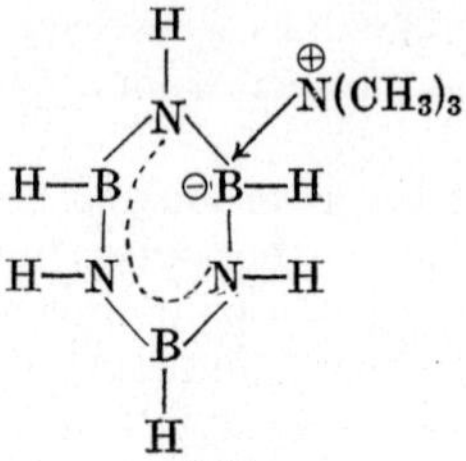

XV

and trimethylamine along with substantial amounts of hydrogen and
trimethylamine-borane as the volatile products. This is consistent
with the proposed structure XV[6, 7].

Most exchange reactions of borazines can be viewed as examples of
nucleophilic substitution. The mechanism is readily interpreted in
terms of an intermediate resulting from the facile change of boron
coordination from three to four caused by the electron deficiency on
the boron atom. This view is supported by the observation that step-
wise replacement of boron substituents is quite possible (see Chapter
III-D). The recently reported iodination of B—H borazines[8] can be
considered as an electrophilic attack of I[⊕] on the hydrogen bonded
to boron. On the other hand, the iodination of decaborane at the 3-boron
atom has been described as an attack by I[⊕] on the boron[9]. However,
in view of the higher electron density in decaborane, this iodination
reaction does not seem to be comparable to the borazine case and final
decision rests upon further results.

---

[1] JONES, R. G., and C. R. KINNEY: J. Amer. chem. Soc. **61**, 1378 (1939).

[2] KINNEY, C. R., and M. T. KOLBEZEN: ibid. **64**, 1584 (1942).

[3] KINNEY, C. R., and C. L. MAHONEY: J. org. Chemistry **8**, 526 (1943).

[4] NIEDENZU, K., D. H. HARRELSON and J. W. DAWSON: Chem. Ber. **94**, 671 (1961).

[5] GERRARD, W., E. F. MOONEY and D. E. PRATT: J. appl. Chem. **13**, 127 (1963).

[6] WEIBRECHT, W. E., and A. W. LAUBENGAYER: Abstr. of Papers, 145th National Meeting of the American Chemical Society, New York 1963, p. 4-N.

[7] WEIBRECHT, W. E.: Ph. D. Thesis, Cornell University, 1964.

[8] MUSZKAT, K. A., and B. KIRSON: Israel J. Chem. **1**, 150 (1963).

[9] LIPSCOMB, W. N.: J. physic. Chem. **61**, 23 (1957).

4. Preparative Examples[1]

B-Trichloro-N-trimethylborazine[2]

A two-liter, three-necked flask is equipped with a stirrer, a water cooled condenser topped with a Dry Ice condenser, and a gas inlet tube which extends to the bottom of the flask. The gas inlet tube is connected with a reservoir of trichloroborane and the equipment is flushed with dry nitrogen. The flask is then charged with one mole (67.5 g.) of anhydrous methylammonium chloride and one liter of dry chlorobenzene and heated to reflux temperature with vigorous stirring. The trichloroborane, 90 ml. (1.1 mole), is then carried by a slow stream of nitrogen into the reaction flask. When all of the trichloroborane has been added, refluxing is continued until the evolution of hydrogen chloride has almost completely ceased; this normally requires about eighteen hours. The reaction mixture is filtered while hot from a minor residue and the solvent removed by vacuum distillation. The crude residue of B-trichloro-N-trimethylborazine can be purified by sublimation or recrystallization from benzene, m.p. 162—164°; the yield is almost quantitative.

N-Trimethylborazine[3]

A 150 cc. flask, equipped with reflux condenser, stirrer and dropping funnel is charged with 4.033 g. of methylammonium chloride and 50 ml. of a 0.476 molar solution of lithium hydroborate in ether is added. After the evolution of hydrogen has ceased, the mixture is refluxed for three hours and the ether is stripped off. The contents of the flask are then heated to 250° for four hours. The product is purified by fractional condensation in high vacuum. The procedure is readily scaled up, permitting the use of normal distillation equipment. The yield is nearly quantitative, b.p. 28° (15 mm.), 133° (760 mm.).

Alternate procedure[4]

A mixture of 10 g. methylammonium chloride and 5.6 g. sodium hydroborate is heated to reflux in triglyme until the evolution of hydrogen has ceased. The borazine is recovered from the solution by distillation under reduced pressure.

B-Tri-p-tolylborazine[5]

A stream of dry ammonia is passed into a solution of 4.3 g. of p-tolydichloroborane in 50 cc. of dry benzene for thirty minutes. The addition of ammonia is continued for one more hour while the reaction mixture is slowly heated to reflux. After being centrifuged, the clear solution is decanted from the precipitated ammonium chloride and the solvent is evaporated. Isopentane is added to the residue and the resultant precipitate is washed with isopentane and recrystallized from a mixture of benzene and isopentane, yielding 1.8 g. of the desired material, m.p. 189—190°.

---

[1] Even if not specifically designated for each preparation, all equipment has to be carefully dried and working in inert gas atmosphere (nitrogen, argon) is recommended. All solvents have to be absolutely anhydrous and proper protection of the reaction vessel contents against exposure to moisture is essential for obtaining reasonable yields of products.

[2] RYSCHKEWITSCH, G. E., J. J. HARRIS and H. H. SISLER: J. Amer. chem. Soc. 80, 4515 (1958).

[3] SCHAEFFER, G. W., and E. R. ANDERSON: ibid. 71, 2143 (1949).

[4] HAWORTH, D. T., and L. F. HOHNSTEDT: Chem. and Ind. 1960, 559.

[5] MIKHAILOV, B. M., A. N. BLOKHINA and T. V. KOSTROMA: J. Gen. Chem. USSR. 29, 1483 (1959).

B-Triethyl-N-trimethylborazine[1]

A 250 cc. two-necked flask is charged with 75 ml. of anhydrous ether, 9.0 g. of magnesium turnings and 25 g. of B-trichloro-N-trimethylborazine. A mixture of 40 g. of ethyl bromide and 25 ml. of ether is added slowly through a dropping funnel to the stirred mixture at such a rate that gentle reflux of the contents is effected while the Grignard reaction is proceeding. Upon completion of the addition, the reaction mixture is refluxed for another three to four hours. After cooling to room temperature, the product is filtered and the ether is stripped off. Fractionation of the residue yields 15 g. of the desired material, b.p. 98° (1.8 mm.), m.p. 1—2°.

B-Tris(dimethylamino)borazine[2]

A solution of 20 g. of anhydrous dimethylamine in 200 cc. of dry benzene is placed in a three-necked flask equipped with stirrer, dropping funnel and Dry Ice condenser. A solution of 10 g. of B-trichloroborazine in 200 cc. of dry benzene is added over a period of fifteen minutes. The mixture is stirred for about one hour at ambient temperature, filtered in an inert atmosphere and the benzene is removed in vacuum. The residue is purified by sublimation at 95° (2 mm.). Yield: about 8 g., m.p. 112—113°.

B-Tris(dibutylamino)borazine[3]

A mixture of 19.8 g. of di-n-butylamine and 10.1 g. of B-tris(diethylamino)-borazine is heated to about 150° for two to three hours. Rectification in vacuum yields 10.7 g. of the desired material, b.p. 200° (0.05 mm.).

B-Tris(N,N-dimethylhydrazino)-N-triethylborazine[4]

A solution of 24 g. of anhydrous N,N-dimethylhydrazine in 400 cc. of dry benzene is heated to reflux and a solution of 29.4 g. of B-tris(dimethylamino)-N-triethylborazine in 200 cc. of dry benzene is added drop-wise over a period of two to three hours. The mixture is refluxed for another four hours. The solvent and any excess of N,N-dimethylhydrazine are removed under reduced pressure and the residue distilled in vacuum to yield 26.2 g. of the desired compound, b.p. 140° (3 mm.).

B-Tris(isocyanato)-N-tri-n-propylborazine[5]

Silver cyanate is dried for two hours at 130° and 20.9 g. are added to a solution of 9.65 g. of B-trichloro-N-tri-n-propylborazine in 25 ml. of dry benzene; an exothermic reaction occurs. The reaction mixture is refluxed for one hour. After filtration, the solvent is removed under reduced pressure and fractional distillation of the residue yields 8.76 g. of the material, b.p. 118° (0.003 mm.).

B-Trimethoxy-N-triethylborazine[4]

A solution of 29.4 g. of B-tris(diethylamino)-N-triethylborazine in 250 cc. of dry benzene is heated to reflux and a mixture of 9.6 g. of absolute methyl alcohol and 100 cc. of dry benzene is added drop-wise over a period of two hours. A slow stream of dry nitrogen is passed through the reaction vessel and refluxing is continued for several hours. After evaporation of the solvent, the residue is distilled under vacuum to yield 12 g. of material, b.p. 93—96° (3 mm.).

[1] RYSCHKEWITSCH, G. E., J. J. HARRIS and H. H. SISLER: J. Amer. chem. Soc. 80, 4515 (1958).

[2] NIEDENZU, K., and J. W. DAWSON: ibid. 81, 3561 (1959).

[3] GERRARD, W., H. R. HUDSON and E. F. MOONEY: J. chem. Soc. (London) 1962, 113.

[4] NIEDENZU, K., D. H. HARRELSON and J. W. DAWSON: Chem. Ber. 94, 671 (1961).

[5] LAPPERT, M. F., and H. PYSZORA: J. chem. Soc. (London) 1963, 1744.

5. Physical Chemistry of Borazines

a. General

Most of the physicochemical investigations of borazines have been concerned with the elucidation of the molecular structure and electronic states of the inorganic heterocycle. For example, LONSDALE and TOOR found that the diamagnetic anisotropy of B-trichloroborazine is about 40% that of 1,3,5-tribromobenzene or 30% that of benzene[1]. This, of course, cannot be taken as a measure of the diamagnetic anisotropy of borazine itself, but it seems probable that the reduction in molecular diamagnetic anisotropy might be a possible measure of the degree of aromatic character. Resistance measurements of B-trimethyl-N-trimethylborazine and of B-triphenylborazine were compared with those of the corresponding benzene derivatives[2]. Although the mechanism of electronic conduction in organic solids is hardly understood as yet, a favorable correlation between the borazines and benzenes was reported.

The structure of borazines suggests that they might form π-bonded intermolecular charge-transfer complexes with other suitable molecules, namely electron acceptors. Indeed CHAMPION et al. found[3] that a 1:1 molar solution of B-trimethyl-N-trimethylborazine and tetracyanoethylene in chloroform gives a broad featureless band in the ultraviolet spectrum, $\Sigma_{max.} = 2000—500$; $\lambda_{max.} = 461$. Such bands are typical of a charge-transfer complex and, since neither of the components alone has this characteristic, they concluded that a complex was formed. A comparison of the frequency of this complex with the frequencies of a mixture of tetracyanoethylene and other donor molecules of known ionization potential (Ip) suggests that the borazine has an $Ip = 8.5$ ev, if a linear relationship is valid. Mixtures of B-trimethyl-N-trimethylborazine with 2,3-dicyano-p-benzoquinone also exhibit a charge-transfer band[4]. The experimentally determined value of $Ip = 8.6$ ev is in close agreement with the earlier work and suggests that the observed optical absorptions are the result of intermolecular charge-transfer complex formation. A report on the interaction of some methylated borazines with a number of acceptor molecules, for example iodine, picric acid, p-benzoquinone and chloranil, seems to support the existence of charge-transfer complexes. As in the cases described above, new bands were observed in the ultraviolet spectra of the reaction products in the spectral region in which charge transfer bands for aromatic systems occur[5]. However, the interpretation of such data has to be approached with extreme caution. It still appears doubtful if the formation of molecular complexes is indicative of the aromatic nature of borazines; a non-aromatic formulation of borazine, i.e. a structure corresponding to a cyclic trimer, would be expected to act as a Lewis base, i.e. complex formation could be anticipated.

[1] LONSDALE, K., and E. W. TOOR: Acta crystallogr. (Copenhagen) 12, 1048 (1959).
[2] OKAMATO, Y., and A. J. GORDON: Chem. and Ind. 1963, 528.
[3] CHAMPION, N. G. S., R. FOSTER and R. K. MACKIE: J. chem. Soc. (London) 1961, 5060.
[4] FOSTER, R.: Nature 195, 490 (1962).
[5] MELLON JR., E. K., and J. J. LAGOWSKI: ibid. 199, 997 (1963).

In a nuclear quadrupole resonance study[1], the quadrupole frequencies of ^{35}Cl in B-trichloroborazine were recorded at 19.937 and 19.639 $\pm$ 0.01 Mc at liquid nitrogen temperature and at room temperature respectively. Since the X-ray crystal analysis of B-trichloroborazine[2] had shown that all chlorine atoms are equivalent in the crystal lattice, only one resonance was to be expected. The B—Cl distance of 1.75 Å in the borazine is nearly identical with that in trichloroborane. Therefore, atomic environments about chlorine atoms in both compounds are probably very similar as is evidenced by a ^{35}Cl resonance in trichloroborane at 21.6 Mc. The difference might then arise from the partial aromatic character of a borazine ring, since KEKULE resonance structures will increase the ionic character of the B—Cl bonds. Accordingly, the p-electron defect on chlorine atoms decreases, leading to a decrease in the frequency of quadrupole resonance.

The aromatic nature of the borazine ring has been cited above (see Chapter III-B 2). On analysis of all data on hand, it seems reasonably certain that the borazine system has some aromatic character. However, it is important to recognize that this does not imply a structure involving positively charged nitrogen and negative boron atoms. True, the π-electron density at boron is about 0.478 electrons compared to 1.522 on nitrogen[3]. However, this distribution is coupled with σ-electron densities in which σ-electron charge has been partially transferred from boron to nitrogen. Consequently, one is justified in drawing a parallel between borazine and the classical benzene structure providing one recognizes the existence of "neutral" boron and nitrogen atoms grouped in the three canonical structures described below. PARISER-PARR-POPLE calculations[3] demonstrate that input parameters based on the $B^{\ominus}-N^{\oplus}$ model reflect a π-electron charge distribution of 0.045 on boron and 1.955 on nitrogen. This is clearly in contrast to the experimental results. For example, quadrupole coupling constants of hexagonal boron nitride suggest donation of 0.45 electrons to the boron in this planar system of trigonal bonds to neighboring nitrogen atoms[4]. This value is in excellent agreement with the data of LCAO molecular orbital calculations based on a neutral borazine molecule. Hence, borazines can be formulated as follows:

$$(III-33)$$

---

[1] NAKAMURA, D., H. WATANABE and M. KUBO: Bull. chem. Soc. Japan **34**, 142 (1961).

[2] COURSEN, D. L., and J. L. HOARD: J. Amer. chem. Soc. **74**, 1742 (1952).

[3] CHALVET, O., R. DAUDEL and J. J. KAUFMAN: Advances in Chemistry **42**, 251 (1964).

[4] SILVER, A. H., and P. J. BRAY: J. chem. Physics **32**, 288 (1960).

b. The Spectra of Borazines

On the basis of the ultraviolet spectra from some phenylated bora-zines Becher and Frick concluded[1] that the free electron pair of the nitrogen participates in the boron-nitrogen bond, i.e. benzene-like resonance structures can be formulated. In B-trimethyl-N-triphenyl-borazine there are no reciprocal effects between the π-electrons of the B—N bond and the aromatic substituent. This can easily be explained by steric influences since the rings are most likely not coplanar.

Ring and Koski[2] concluded that in B-trimethylborazine, there is as much as 68% participation of double-bonded structures. Similar results were reported by Rector, Schaeffer and Platt[3] who, on the basis of ultraviolet data, arrived at a value of 64% for this same compound in agreement with later work by Watanabe, Ito and Kubo[4]. This figure compares favorably with a value of about 47% for borazine itself and it was estimated that, in B-trichloroborazine, the double-bond character is as high as 79%.

As was pointed out previously, it is difficult to evaluate the aromatic character of substituted borazines. A simplified approach involves an evaluation of the bond orders within a given series of borazines and a study of so-called aromatic reactions of borazines.

On the assumption that the asymmetric B—N infrared absorption can be directly related to the bond order, the following series of decreasing aromaticity was established for a number of substituted bor-azines[5] (Table III-12).

However, since N-methylation usually increases the donor prop-erties of nitrogen, it is obvious from these values that coupling effects, for instance, between B—N and N—H vibrations play an important role in the relative values of ν_{BN}. Nevertheless, the position of the B—N stretch might be used to provide a rough evaluation of the aromatic char-acter of a borazine. Therefore, in-

Table III-12

Infrared Absorptions of the B—N Stretch of Some Symmetrically Substituted Borazines

B-substituents	N-substituents	ν_{BN}, cm.$^{-1}$
phenyl	hydrogen	1472
hydrogen	hydrogen	1465
chlorine	hydrogen	1445
phenyl	methyl	1405
hydrogen	phenyl	1401
chlorine	methyl	1392
methyl	phenyl	1380
chlorine	phenyl	1373

frared and Raman spectroscopy have so far played a most important part in elucidating borazine structures. However, assignments have to be made with extreme caution. For instance, due to the ease of synthesizing B-tri-chloro-N-trimethyl-borazine, this latter compound is widely used for preparative studies. However, assignment of even major absorptions of

[1] Becher, H. J., and S. Frick: Z. physik. Chem. (NF) 12, 241 (1957).
[2] Ring, M. A., and W. S. Koski: J. chem. Physics 35, 381 (1961).
[3] Rector, C. W., G. W. Schaeffer and J. R. Platt: ibid. 17, 460 (1949).
[4] Watanabe, H., K. Ito and M. Kubo: J. Amer. chem. Soc. 82, 3294 (1960).
[5] Becher, H. J., and S. Frick: Z. anorg. allg. Chem. 295, 83 (1958).

this compound have long been uncertain. KUBO and coworkers[1] assigned a band at 1087 cm.$^{-1}$ to the N—CH$_3$ rocking vibration and a doublet at 915/980 cm.$^{-1}$ to a B—Cl stretch. Although reversed by later authors[2], the most recent work[3] confirms the findings of KUBO. Similarly, KUBO[1, 4] assigned the strongest band of N-methylated borazines to a B—N ring frequency in agreement with studies by BECHER and FRICK[5]. That assignment was finally confirmed by a spectroscopic study of (—BR—NCD$_3$—)$_3$ (R = Cl, NH$_2$)[6].

The distribution of potential energy among various internal coordinates was calculated for each normal mode of vibration of borazine and B-trichloroborazine. On the basis of such calculations, the assignment of major infrared absorption to fundamentals were effected in order to elucidate the coupling between various stretchings and deformations of bond distances and valence angles, respectively. The assignments of major infrared absorptions of B-trichloroborazine[1] are listed in Table III-13.

Table III-13. *Assignment of Major Infrared Absorptions of B-Trichloroborazine*

absorption (cm.$^{-1}$)	assignment
3442	N—H stretching
1452 (sh) ⎫ 1442 ⎭	B—N ring vibration
1037 (sh) ⎱ 1032 ⎰	⎰ N—H bending in the molecular plane ⎱ coupled with B—Cl stretching
744	B—Cl and N—H bending
706	B—Cl and N—H stretching

The shoulder at 1452 typically denotes an effect of the ^{10}B isotope. In naturally abundant boron compounds, the isotopic ratio is ^{11}B:^{10}B = 81:19. If the isotopic shift is calculated by the perturbation method, a theoretical value of about 14 cm.$^{-1}$ is obtained, in close agreement with the experimental observations.

Assignments of B-trichloro-N-trimethylborazine and B-trimethyl-N-trimethylborazine were made in an analogous manner. On this basis, the observation of strong B—N stretching frequencies was used to assign borazine structures for a number of derivatives listed in Table III-14.

[1] WATANABE, H., M. NARISADA, T. NAKAGAWA and M. KUBO: Spectrochim. Acta **16**, 78 (1960).

[2] BUTCHER, J. M., W. GERRARD, E. F. MOONEY, R. A. ROTHENBURY and H. A. WILLIS: ibid. **18**, 1487 (1962).

[3] GUTMANN, V., A. MELLER and R. SCHLEGEL: Mh. Chem. **94**, 1071 (1963).

[4] WATANABE, H., Y. KURODA and M. KUBO: Spectrochim. Acta (London) **17**, 454 (1961).

[5] BECHER, H. J., and S. FRICK: Z. anorg. allg. Chem. **295**, 83 (1959).

[6] MELLER, A., and R. SCHLEGEL: Mh. Chem. **95**, 382 (1964).

However, as cited earlier, the simultaneous observation of the B—N stretching frequency in conjunction with a B—N out-of-plane vibration seems to have a general diagnostic value. In recent years, a variety of compounds containing a covalent boron-nitrogen linkage have become available and their spectra indicate that strong absorption in the 1350—1500 cm.$^{-1}$ range is not necessarily diagnostic of a borazine ring. Attention to a medium-intensity doublet in the 720 cm.$^{-1}$ region of the spectra of B-aminoborazines and B-alkoxy-borazines was first drawn by LAPPERT[1]. Later, an attempt was made to consider these out-of-plane vibrations in conjunction with the B—N stretch for determining the existence of a borazine ring structure[2]. Despite certain limitations, an appreciable diagnostic value of such considerations was demonstrated.

Table III-14

B—N Stretch of Some Borazines

B-substituents	N-substituents	ν_{B-N}(cm.$^{-1}$)
CH₃	C₆H₅	1376
C₆H₅	C₆H₅	1368
Cl	C₆H₅	1378
C₄H₉	C₆H₅	1382
C₂H₅	C₆H₅	1387
Cl	CH₃	1392
C₂H₅	C₂H₅	1417
Cl	C₄H₉	1418
H	D	1436
Cl	C₂H₅	1442
Cl	H	1442

The proton magnetic resonance spectra of several borazines have been recorded by ITO, WATANABE and KUBO[3]. The chemical shifts of B—H and N—H protons of variously substituted borazines are very close to each other (Table III-15).

Table III-15. *Chemical Shifts of Single Protons in Some Borazines*

Compound	Chemical Shifts	
	B—H proton	N—H proton
N-trimethylborazine, (—BH—NCH₃—)₃	—3.05	
B-trichloroborazine, (—BCl—NH—)₃		—3.93
borazine, (—BH—NH—)₃	—3.07	—4.05

From the relationship of the chemical shifts of methyl and methylene resonances in some N-triethylborazines, the electronegativity of the ring nitrogen was calculated (using the DAILEY-SHOOLERY formula) as:

$$E = 0.695\,(T_{CH_3} - T_{CH_2}) + 1.71 \qquad (III\text{-}32)$$

The relative electronegativities E are shown in Table III-16.

These data[4] support infrared spectroscopic evidence[5] relating the effect of boron substituents to the degree of boron-nitrogen double

[1] LAPPERT, M. F.: Proc. chem. Soc. (London) **1959,** 59.

[2] BEYER, H., J. B. HYNES, H. JENNE and K. NIEDENZU: Advances in Chemistry **42,** 266 (1964).

[3] ITO, K., H. WATANABE and M. KUBO: Bull. chem. Soc. Japan **33,** 1580 (1960).

[4] MOONEY, E. F.: Spectrochim. Acta **18,** 1355 (1962).

[5] BUTCHER, I. M., W. GERRARD, E. F. MOONEY, R. A. ROTHENBURY and H. A. WILLIS: ibid. **18,** 1487 (1962).

bond character in borazines; it appears to increase in the order
Br $<$ Cl $<$ C_6H_5.

It is of interest to note that the proton chemical shift of arylated
borazines is recorded at a relatively high field[1, 2]. On the basis of ultraviolet data, BECHER and FRICK[3]
had originally suggested that the
orientation of aromatic nuclei is
perpendicular to the plane of the
borazine ring in such substituted
borazines. The proton magnetic
resonance data appear to confirm
these findings, since the shielding
effect of the aromatic ring currents
on methylene groups held above
the plane of the ring is fully established[4]. In conclusion, it should be
noted that borazines do not produce electron-spin resonance spectra[5].

Table III-16.

*Relative Electronegativities of Nitrogen
in Several Substituted Borazines*

Compound	E
$(-BCl-NC_2H_5-)_3$	3.46
$(-BBr-NC_2H_5-)_3$	3.56
$(-BC_6H_5-NC_2H_5-)_3$	3.24

D. Unsymmetrically Substituted Borazines

One of the most intriguing problems in preparative borazine chemistry involves the synthesis of unsymmetrically substituted borazines
and a study of their reactions. Only if the difficulties inherent in their
preparation are overcome can detailed investigations of substitution
effects at the borazine ring be effected.

Only a few borazine derivatives which are unsymmetrically substituted with respect to one or both of the skeletal atoms have been described.
2,4-Dibromoborazine has long been known[6], and the formation of unsymmetrically methylated borazines was reported even earlier[7, 8], but
these compounds are very difficult to prepare. Partial replacement of
boron-attached halogen by organic groups via organometallics has
been reported[9-11] but the yield of product is quite small. The steric
influence of the N-substituents in the course of B-alkylation via GRIGNARD
reagents has been discussed and the reaction of equimolar amounts
of B-trichloro-N-triphenyborazine with n-butyl GRIGNARD was claimed
to provide the unsymmetrical product in reasonable yield[11]. In contrast,
TOENISKOETTER and HALL[12] state that partial alkylation of B-trichloro-

[1] MOONEY, E. F.: Spectrochim. Acta **18**, 1355 (1962).
[2] ITO, K., H. WATANABE and M. KUBO: J. chem. Physics **32**, 947 (1959).
[3] BECHER, H. J., and S. FRICK: Z. physik. Chem. (NF) **12**, 241 (1957).
[4] WAUGH, J. S., and R. W. FESSENDEN: J. Amer. chem. Soc. **79**, 846 (1957).
[5] CHAPMAN, D., S. H. GLARUM and A. G. MASSEY: J. chem Soc. (London) **1963**, 3140.
[6] WIBERG, E., and A. BOLZ: Ber. dtsch. chem. Ges. **73**, 209 (1940).
[7] SCHLESINGER, H. I., L. HORVITZ and A. B. BURG: J. Amer. chem. Soc. **58**, 409 (1936).
[8] SCHLESINGER, H. I., D. M. RITTER and A. B. BURG: ibid. **60**, 1296 (1938).
[9] RYSCHKEWITSCH, G. E., J. J. HARRIS and H. H. SISLER: ibid. **80**, 4515 (1958).
[10] SMALLEY, J. H., and S. F. STAFIEJ: ibid. **81**, 582 (1959).
[11] MELLER, A.: Mh. Chem. **94**, 183 (1963).
[12] TOENISKOETTER, R. H., and F. R. HALL: Inorg. Chem. **2**, 29 (1963).

borazines is fairly selective only if the nitrogen is substituted with methyl groups. Indeed, reasonable yields of unsymmetrical products have been reported by Gutmann, Meller and Schlegel when B-trichloro-N-trimethylborazine is treated with n-butylmagnesium halide[1]. Alkylation of B—H borazines with insufficient amounts of organometallics has been studied and was found to give a random distribution of products.

It is of interest to note that 1,3,5-trimethyl-2,4-dimethylborazine reacts with hydrogen halide to yield 1,3,5,2,4-pentamethyl-6-chloroborazine. It is very probable that, initially, a 1:3 addition product was formed which, on pyrolysis, afforded the monohalogenated derivative[2].

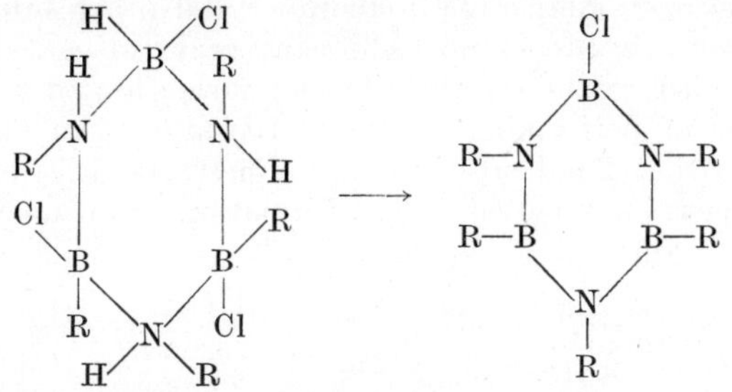

$$(\text{III-34})$$

This method appears to be most facile for the preparation of B-mono- and dihalogenated substituted borazines, since isolation and separation of products is more easily achieved in this manner than from an organo-metallic synthesis with B-trihalogenoborazines.

The partial fluorination of B-trichloroborazine[3] has not yet been developed for large scale preparation. However, by recognizing the steric factors that seem to predominate in displacement reactions of the three-coordinate boron compounds[4], Lappert and coworkers have suggested a new and interesting approach to the problem. The preparation of some unsymmetrical B-aminoborazines and B-alkoxyborazines[5] and some unsymmetrical B-isocyanatoborazines and B-isothiocyanato-borazines[6] has been effected in good yield. For example, B-tris(diethyl-amino)-N-triethylborazine and ethylamine were reacted in equimolar portions to give a nearly quantitative yield of the unsymmetrical compound.

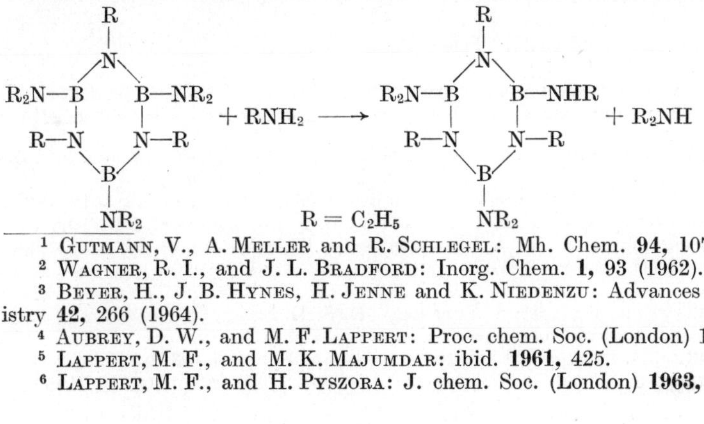

$$(\text{III-35})$$

$R = C_2H_5$

[1] Gutmann, V., A. Meller and R. Schlegel: Mh. Chem. 94, 1071 (1963).
[2] Wagner, R. I., and J. L. Bradford: Inorg. Chem. 1, 93 (1962).
[3] Beyer, H., J. B. Hynes, H. Jenne and K. Niedenzu: Advances in Chemistry 42, 266 (1964).
[4] Aubrey, D. W., and M. F. Lappert: Proc. chem. Soc. (London) 1960, 148.
[5] Lappert, M. F., and M. K. Majumdar: ibid. 1961, 425.
[6] Lappert, M. F., and H. Pyszora: J. chem. Soc. (London) 1963, 1744.

In an analogous manner, B-trichloroborazines were reacted with metal cyanates and thiocyanates to yield B-halogeno-pseudohalogenoborazines.

On the basis of this relatively limited information, it is possible to conclude that a partial displacement reaction at a borazine nucleus might best be achieved where similar groups are involved; for example, replacing halogen by pseudohalogen and amino groups through other amines. Since B-attached amino and pseudohalogeno groups on a borazine ring are quite useful for subsequent reactions, this method deserves further attention. This is particularly true since a number of exchange reactions of substituted borazines with trimethylborane[1], trihalogenoboranes[2], aminodimethylborane[3] and other substituted borazines[4, 5] apparently are of no significant preparative value.

A report also exists on the unusual formation of a B—H bond in the process of performing a BROWN-LAUBENGAYER synthesis with bulky amines, i.e. 2,6-disubstituted anilines[6]. In this case, thermal dehydrohalogenation resulted in the formation of an unsymmetrically substituted borazine.

$$3\ RNH_2 \cdot BCl_3 \longrightarrow \qquad\qquad\qquad\qquad\qquad\qquad (III\text{-}36)$$

However, since details are unavailable at this time, it is difficult to evaluate the inherent possibilities of this method for preparative purposes.

Unsymmetrical replacement on the nitrogen of a borazine ring has been reported only recently[7]. By treatment of B-trimethylborazine with an equimolar quantity of methyllithium, B-trimethyl-N-lithioborazine was obtained. Subsequent treatment of this product with alkyl halide afforded B-trimethyl-N-monoalkylborazine. In an analogous procedure, pentaalkylborazine was synthesized, but the inherent possibilities of this preparative method have not yet been explored.

2-Methyl-1,3,5-triphenylborazine[8]

A GRIGNARD solution, prepared from 10.95 g. of methyl iodide in 70 ml. of ether, is added to a stirred solution of 21.6 g. of N-triphenylborazine in 300 cc. of dry

[1] SCHLESINGER, H. I., D. M. RITTER and A. B. BURG: J. Amer. chem. Soc. 60 1296 (1938).

[2] SCHAEFFER, G. W., R. SCHAEFFER and H. I. SCHLESINGER: ibid. 73, 1612 (1951).

[3] SCHLESINGER, H. I., L. HORVITZ and A. B. BURG: ibid. 58, 409 (1936).

[4] RYSCHKEWITSCH, G. E., J. J. HARRIS and H. H. SISLER: ibid. 80, 4515 (1958).

[5] NEWSOM, H. C., W. G. WOODS and A. L. McCLOSKEY: Inorg. Chem. 2, 36 (1963).

[6] BARTLETT, R. K., H. S. TURNER, R. J. WARNE, M. A. YOUNG and W. S. McDONALD: Proc. chem. Soc. (London) 1962, 153.

[7] WAGNER, R. I., and J. L. BRADFORD: Inorg. Chem. 1, 93 (1962).

[8] SMALLEY, J. H., and S. F. STAFIEJ: J. Amer. chem. Soc. 81, 582 (1959).

ether over an hour. The mixture is stirred at room temperature for two hours and titrated with a saturated aqueous solution of ammonium chloride to the point where magnesium salts settle rapidly from solution. The ether solution is decanted from the precipitated salts and evaporated to dryness. Recrystallization of the solid product from hexane or petrol ether (b.p. 30—60°) yields about 20 g. of pure product, m.p. 140—142°.

2,4-Di-n-butyl-6-chloroborazine[1]

A GRIGNARD solution is prepared from 29 g. of n-butyl bromide and 6 g. magnesium turnings in 500 cc. of ether. After filtration, it is added to a solution of 36.8 g. of B-trichloroborazine in 800 cc. of dry benzene over a period of two hours. The reaction mixture is refluxed for thirty minutes and the solvents stripped off under reduced pressure. The stripping is interrupted once or twice to filter off the precipitated magnesium salts. Vacuum distillation of the final residue yields 14 g. of the desired compound, b.p. 91°; it is quite unstable and disproportionation products appear after a few hours standing.

1,2,4,6-Tetramethylborazine[2]

A quantity, 1.35 g. of B-trimethylborazine is reacted with 11.6 ml. of a 0.95 molar ethereal solution of methyllithium in a 30 ml. glass bomb tube closed with a rubber septum. After cessation of methane evolution, a 3.40 ml. aliquot of a 10 ml. ethereal solution containing 4.58 g. of methyl iodide (11.02 mmoles), is added from a syringe. The sealed tube is heated at 80° for four hours, cooled and opened. The ether is removed on the vacuum line and the residue extracted with eight 2-ml. portions of n-pentane. After evaporation of the pentane, the product is fractionated by vapor phase chromatography. The yield is about 40% of product, b.p. 129—131°.

2,4-Dichloro-1,3,5,6-tetramethylborazine[3]

Fifty g. of B-trichloro-N-trimethylborazine and 250 cc. of dry ether are placed in a 500 cc. flask equipped with a stirrer, reflux condenser and dropping funnel. The borazine solution is stirred rapidly and 54 ml. of a 4.1 molar solution of methylmagnesium bromide in diethyl ether (0.22 mole) is added at a rate which maintains gentle reflux. The mixture is stirred for an additional two hours and set aside for eighteen hours. The solution is filtered through glass wool and the clear yellow filtrate is evaporated to dryness. Recrystallization of the residue from 1,2-dimethoxyethane gives 37 g. of the compound, m.p. 143—145°.

E. Polynucleated Borazines

In STOCK's original work on borazine[4] it was noted that, in the liquid state, borazine tends to dehydrogenate at ambient temperatures and to form polymeric substances intermediate between $B_3H_3N_3H_3$ and $(BNH)_x$. However, little additional effort has been made to investigate the nature of these products[5]. Presumably they consist of borazine rings linked together and they may be related to the higher homologs of benzene, just as borazine has been related to benzene itself. Closely related to these polymeric decomposition products of borazine might be the nonvolatile materials obained as byproducts in the BROWN-LAUBEN-

[1] MELLER, A.: Mh. Chem. 94, 183 (1963).

[2] WAGNER, R. I., and J. L. BRADFORD: Inorg. Chem. 1, 93 (1962).

[3] TOENISKOETTER, R. H., and F. R. HALL: ibid. 2, 29 (1963).

[4] STOCK, A.: Hydrides of Boron and Silicon, Cornell University Press, Ithaca, N.Y., 1933.

[5] SCHAEFFER, R., M. STEINDLER, L. F. HOHNSTEDT, H. S. SMITH JR., L. B. EDDY and H. I. SCHLESINGER: J. Amer. chem. Soc. 76, 3303 (1954).

GAYER synthesis of B-trichloroborazine[1]. However, only during the last three or four years have there been some reports published concerning the polynucleated borazines, i.e. substances in which two or more borazines are fused, linked either directly or by bridges formed by other atoms or atom groupings.

LAUBENGAYER and coworkers[2] studied the thermal decomposition of borazine vapor at temperatures between 340° and 440°. The pyrolysis is kinetically first order with respect to borazine and provides a nonvolatile solid material of the approximate composition $BNH_{0.8}$. Besides hydrogen, a number of volatile materials containing boron and nitrogen are obtained. Two of the products, volatile in high vacuum at room temperature, were identified as B—N analogs of naphthalene (XV) and diphenyl (XVI). A third fraction was tentatively identified as 2,4-diaminoborazine. This last identification was confirmed by an independent

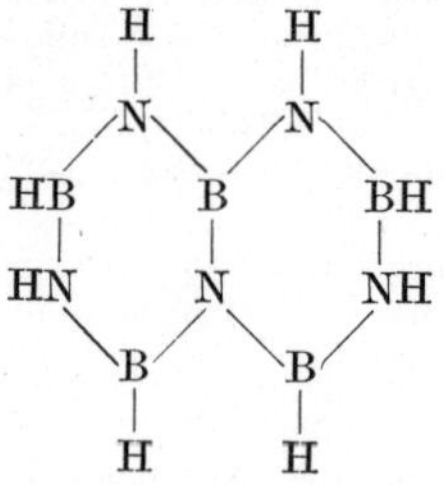

m.p. = 27—30°
XV

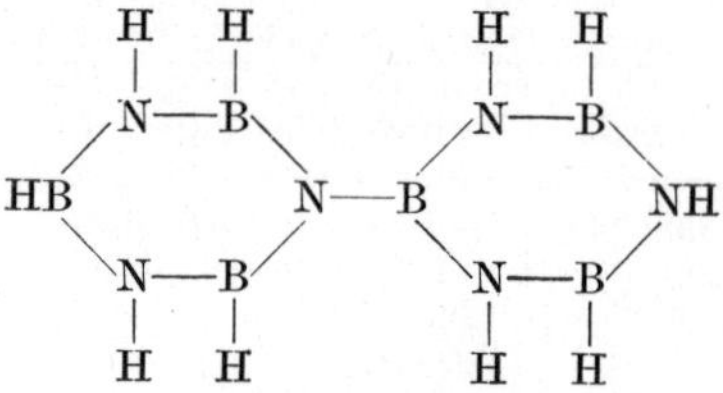

m.p. = 59—60°
XVI

synthesis of this relatively unstable material through dimethylaminolysis of 2,4-dibromoborazine and subsequent transamination of the resultant product with anhydrous ammonia[3]. Spectral features of the 2,4-diaminoborazine prepared in this manner are identical with those reported by LAUBENGAYER for the intermediate in the pyrolysis of borazine. Mass spectroscopic analysis of the borazine pyrolysis products which are nonvolatile at room temperature but volatile at 90° seems to indicate that they may be amino derivatives of XV and XVI or a fused tricyclic system of the composition $B_6N_7H_9$. Infrared spectroscopic data of the nonvolatile residue seems to support their structure as being of a highly condensed boron-nitrogen framework approaching that of boron nitride but still containing N—H and B—H bonds.

It is of interest to note that the thermal decomposition of borazine is not wholly comparable to that of benzene. Intermediates in the thermal decomposition of benzene are biphenyl, terphenyl et al. but not naphthalene. This indicates that intermolecular dehydrogenation is the important mechanism in the pyrolysis of benzene. In the case of borazine, however, extensive ring cleavage seems to be as important as dehydrogenation. This is evidenced by the formation of 2,4-diaminoborazine and of borazanaphthalene (XV).

---

[1] BROWN, C. A., and A. W. LAUBENGAYER: J. Amer. chem. Soc. 77, 3699 (1955).
[2] LAUBENGAYER, A. W., P. W. MOEWS JR. and R. F. PORTER: ibid. 83, 1337 (1961).
[3] NIEDENZU, K., and coworkers: unpublished results.

Little is known about the chemistry of the higher borazine homologs. Only the reaction of borazanaphthalene (which has become more readily available by an electrical discharge method[1]) with hydrogen halides and the subsequent pyrolysis of the addition products has been examined in detail[2]. Regardless of the molar ratios of the reactants, adducts of the formula $B_5N_5H_8 \cdot 5\,HX$ are obtained. This can be interpreted in terms of an initial 9,10-addition of one mole of hydrogen halide at the ring juncture bond thereby destroying the resonance stabilization of the system. Consequently, the addition of four more moles of hydrogen halide appears necessary to form a stable product, XVII.

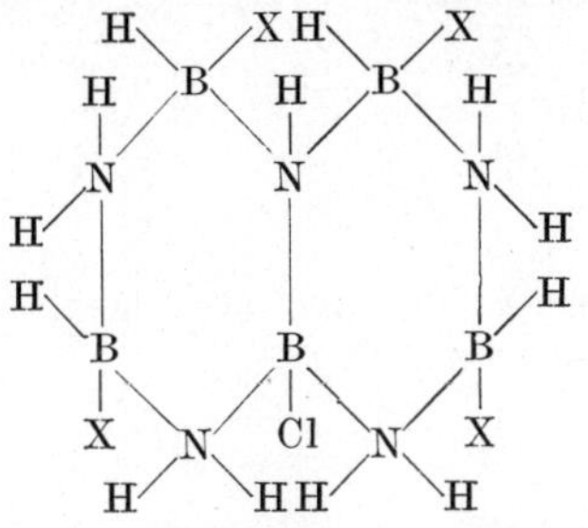

XVII

Thermal decomposition of the adduct XVII results in the formation of hydrogen, hydrogen halide and borazanaphthalene. Small quantities of 2-chloro- and 2,4-dichloroborazine were observed when the adduct was decomposed at 110° in the mass spectrophotometer, thus indicating the existence of secondary reactions. However, the major course of the reaction appears to be closely analogous to that of corresponding borazine adducts.

Polynucleated borazines of the diphenyl type have been studied in greater detail. This work was facilitated by the discovery of N-lithio substituted borazines[3], which were found to be capable of reacting with B-halogenoborazines[4].

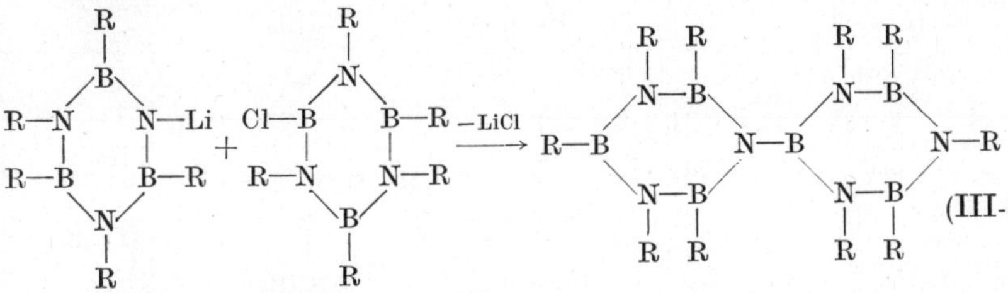

(III-37)

It has also been reported that, during B-alkylation of B-trihalogeno-borazines with GRIGNARD reagents, small amounts of such diborazinyl

[1] LAUBENGAYER, A. W., and O. T. BEACHLEY JR.: Advances in Chemistry 42, 281 (1964).
[2] LAUBENGAYER, A. W.: Personal communication.
[3] WAGNER, R. I., and J. L. BRADFORD: Inorg. Chem. 1, 93 (1962).
[4] WAGNER, R. I., and J. L. BRADFORD: ibid. 1, 99 (1962).

compounds are obtained[1]. It seems rather doubtful, however, that this latter reaction can be used for preparative purposes. In contrast, the condensation of two borazine nuclei to yield diborazinyl derivatives by elimination of lithium chloride (eq. III-37) has been extended to the preparation of borazine polymers (XVIII).

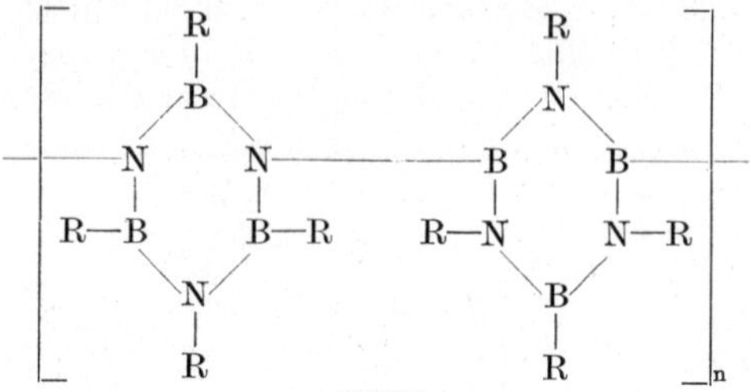

XVIII

Diborazinyl structures (XIX) have also been obtained by pyrolysis of B-tris(dialkylamino)borazines[2].

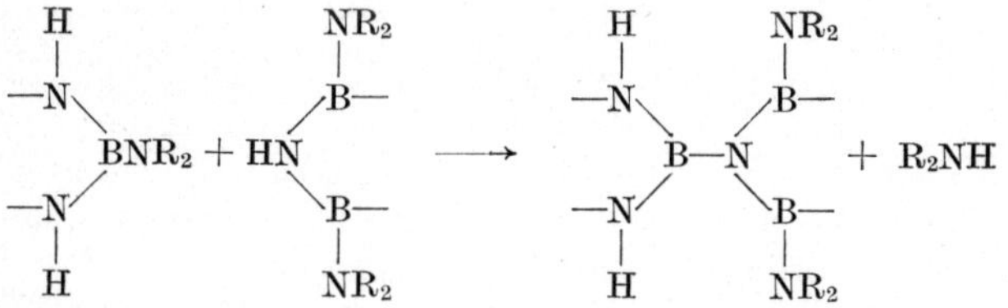

$$+ R_2NH \quad \text{(III-38)}$$

XIX

Diborazinyl structures or polynucleated borazines in which a nitrogen of one borazine ring is linked to nitrogen of a second borazine nucleus are not known. However, GUTMANN and coworkers[3] have recently reported the synthesis of polynucleated borazines with B—B linkages. Such compounds were obtained by reacting B-chlorinated borazines with alkali metals as shown in eq. III-39.

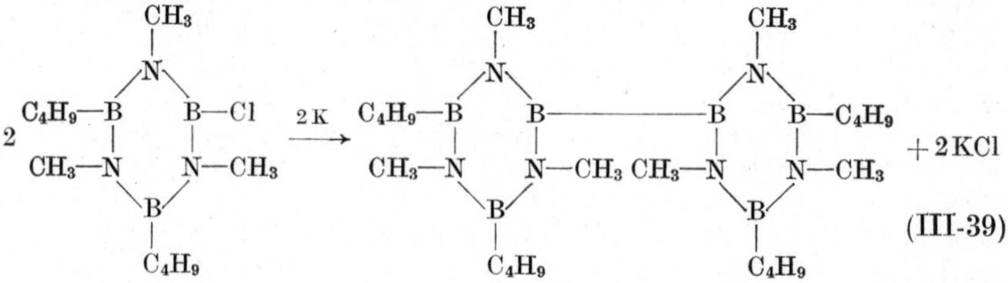

$$+ 2\,KCl$$

$$\text{(III-39)}$$

Another type of polynucleated borazines is formed through linking two or more borazine nuclei by bridges of other atoms or groups of atoms.

---

[1] HARRIS, J. J.: J. org. Chemistry **26**, 2155 (1961).

[2] GERRARD, W., H. R. HUDSON and E. F. MOONEY: J. chem. Soc. (London) **1962**, 113.

[3] GUTMANN, V., A. MELLER and R. SCHLEGEL: Mh. Chem. **95**, 314 (1964).

The first such derivatives were reported by AUBREY and LAPPERT[1, 2]. These authors investigated the thermal decomposition of tris(primary amino)boranes. At temperatures near 200°, B-triamino-N-trisubstituted borazines are obtained; they in turn eliminate amine when heated to 300°.

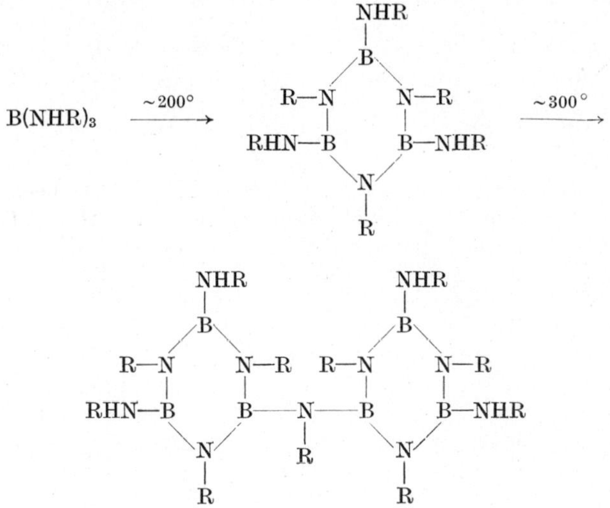

$$
B(NHR)_3 \xrightarrow{\sim 200°}
\begin{array}{c}
NHR \\
| \\
B \\
R-N \qquad N-R \\
| \qquad\qquad | \\
RHN-B \qquad B-NHR \\
N \\
| \\
R
\end{array}
\xrightarrow{\sim 300°}
$$

$$
\begin{array}{cc}
NHR & NHR \\
| & | \\
B & B \\
R-N \quad N-R \quad R-N \quad N-R \\
| \qquad | \qquad | \qquad | \\
RHN-B \quad B-N-B \quad B-NHR \\
N \quad\quad R \quad\quad N \\
| \qquad\qquad\qquad | \\
R \qquad\qquad\qquad R \\
XX
\end{array}
\qquad (III\text{-}40)
$$

The ultimate product of this pyrolysis reaction appears to be highly crosslinked[3]. However, thermal decomposition under carefully controlled conditions yields well characterized bicyclic derivatives of type XX[4]. GUTMANN, MELLER and SCHLEGEL have reported similar results[5]. Bridging of borazine rings by atoms other than nitrogen has also been reported. The first material with an oxygen bridge was accidentally obtained as a by-product from the reaction of a B-trihalogenoborazine derivative with an insufficient amount of Grignard reagent[6]. Borazine polymers with oxygen bridges have been obtained through controlled hydrolysis of mixtures of mono- and dihalogenoborazines[7]; analogous materials are formed on refluxing B-trialkoxyborazines in the presence of acidic catalysts[8]. For example, when B-tris(t-butoxy)-N-trimethyl-borazine was refluxed in the presence of boric acid, isobutene and water were evolved and a compound of type XXI was obtained. Such

[1] LAPPERT, M. F.: Proc. Chem. Soc. (London) 1959, 59.
[2] AUBREY, D. W., and M. F. LAPPERT: J. chem. Soc. (London) 1959, 2927.
[3] TOENISKOETTER, R. H., and F. R. HALL: Inorg. Chem. 2, 29 (1963).
[4] LAPPERT, M. F., and M. K. MAJUMDAR: Proc. chem. Soc. (London) 1961, 425.
[5] GUTMANN, V., A. MELLER and R. SCHLEGEL: Mh. Chem. 94, 1071 (1963).
[6] SEYFERTH, D., and H. P. KÖGLER: J. inorg. nucl. Chem. 15, 99 (1960).
[7] WAGNER, R. I., and J. L. BRADFORD: Inorg. Chem. 1, 99 (1962).
[8] LARCOMBE, B. E., and H. S. TURNER: Chem. and Ind. 1963, 410.

oxygen-bridged borazines, i.e. borazyl oxides, have also been obtained by condensation of B-aminoborazines with dimethylformamide[1].

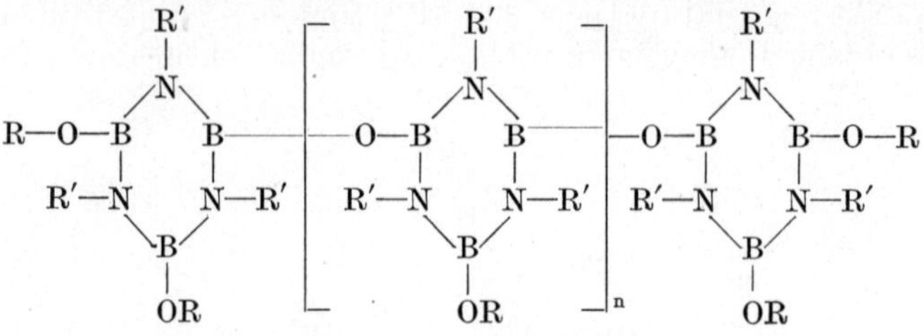

XXI

Carbon-bridged polynucleated borazines, **XXII**, have been prepared by the reaction of B-monohalogenoborazines with the di-GRIGNARD reagent from 1,4-dihalogenobutane[2, 3].

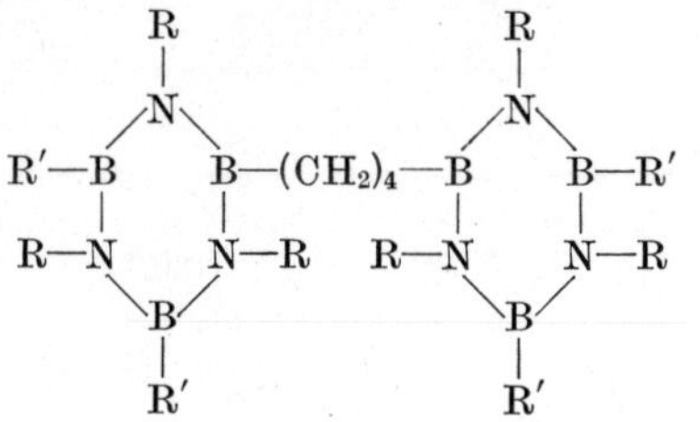

$R = CH_3$
$R' = CH_2[Si(CH_3)_3]$
m.p. $= 93—94°$

XXII

Such compounds appear to be rather unstable toward heat and hydrolysis[2].

Chapter IV

σ-Bonded Cyclic Systems of Boron and Nitrogen

(Other than Borazines)

A. General Remarks

Besides the borazines, several other cyclic systems with no other annular atoms than boron and nitrogen have been discovered during the past two or three years. Examples of such compounds, however, are

---

[1] TOENISKOETTER, R. H., and K. A. KILLIG: J. Amer. chem. Soc. **86**, 690 (1964).
[2] SEYFERTH, D., W. R. FREYER and M. TAKAMIZAWA: Inorg. Chem. **1**, 710 (1962).
[3] HARRIS, J. J., PH. D. THESIS: University of Florida 1958.

still very few. This discussion will be developed with respect to increasing ring size rather than from a historical point of view and the "borine" nomenclature of the systems will be used. To date, derivatives of the following four rings are known:

1. 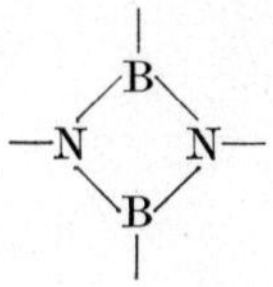
$$
\begin{array}{c}
| \\
\diagup B \diagdown \\
-N \qquad N- \\
\diagdown B \diagup \\
|
\end{array}
$$
 s-diaza-diborine

2. 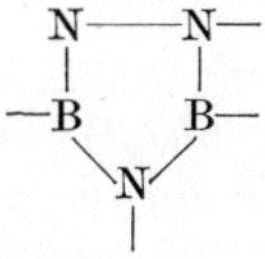
$$
\begin{array}{c}
N\!-\!-\!N- \\
|\qquad\quad\; \\
-B \qquad B- \\
\diagdown N \diagup \\
|
\end{array}
$$
 1,3,4-triaza-2,5-diborine

3. 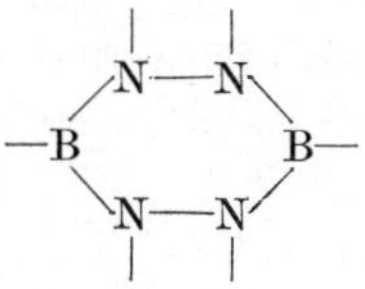
$$
\begin{array}{c}
|\quad\;\; | \\
\diagup N\!-\!N\diagdown \\
-B \qquad\quad B- \\
\diagdown N\!-\!N\diagup \\
|\quad\;\; |
\end{array}
$$
 1,2,4,5-tetraza-3,6-diborine

4. $(-B-N-)_4$ s-tetraza-tetraborine

Numbering, according to the Ring Index, begins with nitrogen. For the s-diaza-diborines, the name 1,3-diaza-2,4-boretane can be found in the literature; derivatives of s-tetraza-tetraborine have been classified as tetrameric borazines, borazocines or borazynes.

B. s-Diaza-diborines

The sole example of a s-diaza-diborine was discovered by LAPPERT and MAJUMDAR[1, 2]. A close relationship exists between this class of compounds, the borazines and the s-tetraza-tetraborines. All three types can be considered as polymeric azaborines, derived from a hypothetical $HN=BH$ in various degrees of polymerisation. Consequently, they are all formed by elimination reactions of aminoboranes. LAPPERT[2] considers the mechanism of formation as a three-step process:

1. An intermolecular condensation of an aminoborane with formation of a diborylamine nucleus, $>\!B-N-B\!<$;

---

[1] LAPPERT, M. F., and M. K. MAJUMDAR: Proc. chem. Soc. (London) 1963, 88.
[2] LAPPERT, M. F., and M. K. MAJUMDAR: Advances in Chemistry 42, 208 (1964).

2. Intramolecular 1,3-nucleophilic rearrangement to yield an aza-borine;

3. Polymerisation of the azaborine.

This procedure can be formulated as follows:

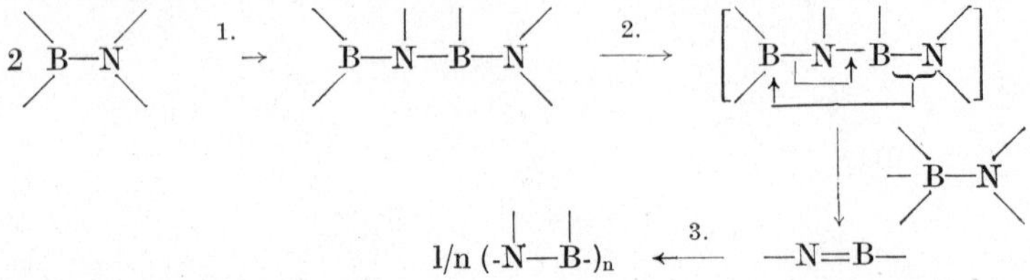

Evidence for this mechanism rests primarily on the isolation of di-borylamines as reaction intermediates. Hence, a systematic study of diborylamines might provide more conclusive evidence for the proposed mechanism. The degree of polymerization might well be controlled by steric factors.

In consonance with the proposed mechanism, one s-diaza-diborine (I) has been obtained by the thermal degradation of either tris(t-butyl-amino)borane or bis(di-t-butylaminoboryl)-t-butylamine.

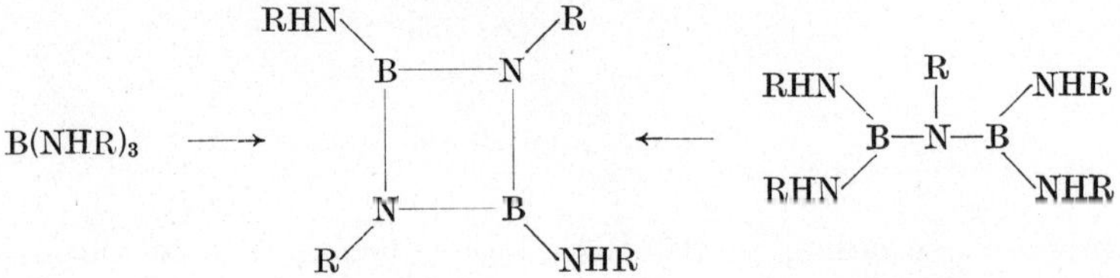

The s-diaza-diborine system is of interest because it may be formally regarded as isoconjugate with cyclobutadiene. The instability of the latter has been explained in terms of molecular orbital theory[1]. The diamagnetism of I implies, however, that electron delocalization within the ring is not significant. In other words, if π-bonding exists in this system, (which can reasonably be assumed on the evidence accumulated for other σ-bonded boron-nitrogen molecules) it will be largely exo-cyclic; thus such bonding might be responsible for the stability of the product.

C. 1,3,4-Triaza-2,5-Diborines

Very little information is available on the 1,3,4-triaza-2,5-diborine system (II). NIEDENZU[2, 3] described the controlled hydrazinolysis of

---

[1] LONGUET-HIGGINS, H. C., and L. E. ORGEL: J. chem Soc. (London) 1956, 1969.

[2] NIEDENZU, K.: Symposium on "Current Trends in Organometallic Chemistry", Cincinnati, Ohio, USA., 1963.

[3] NIEDENZU, K., H. JENNE and P. FRITZ: Angew. Chem. 76, 535 (1964).

bis(dimethylaminophenylboryl)amine and Nöth[1] reported the aminolysis of a 1,2,4,5-tetraza-3,6-diborine. Both reactions lead to the 1,3,4-triaza-2,5-diborine structure:

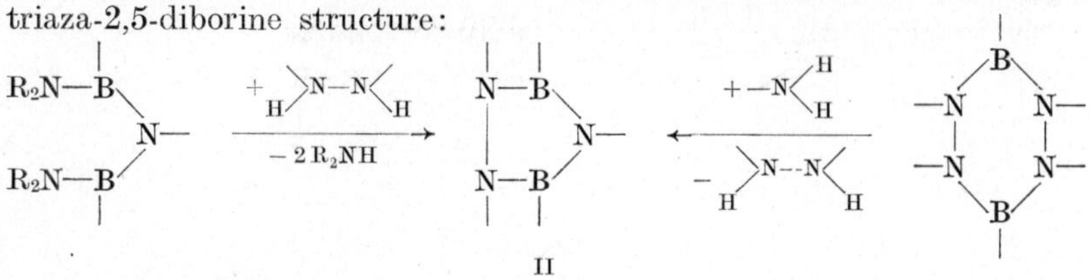

II

Compounds of type II are extremely sensitive to hydrolysis. Existing evidence suggests that it is the B—N—B linkage that is first attacked by water, but the residual hydrazinoborane linkage is also unstable. Compounds of type II appear to be relatively stable toward pyrolysis. Their stability may be the result of a substantial degree of resonance stabilization in the ring. This postulation is confirmed by the observation of a B—N stretching frequency in the 1400 cm.$^{-1}$ region of the infrared spectrum, which region is normally associated with a high degree of B—N bond order.

D. 1,2,4,5-Tetraza-3,6-Diborines

Derivatives of this cyclic boron-nitrogen system have been recognized earlier (see Chapter II-G). The first preparation to be reported was that of 1,2,4,5-tetrahydro-3,6-diphenyl-1,2,4,5-tetraza-3,6-diborine, III[2, 3] obtained by the hydrazinolysis of bis(dimethylamino)phenylborane:

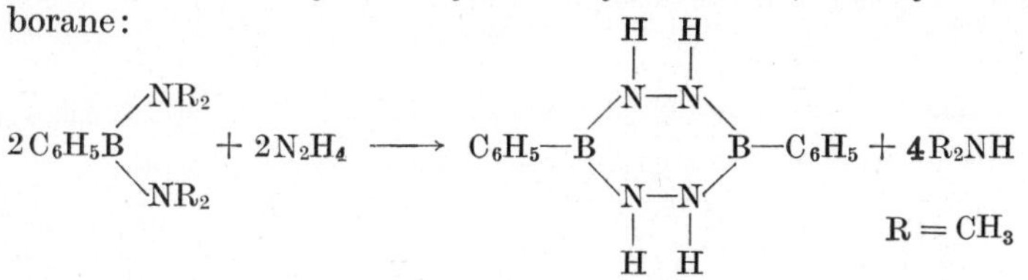

$$2\,C_6H_5B \begin{matrix} \diagup NR_2 \\ \diagdown NR_2 \end{matrix} + 2\,N_2H_4 \longrightarrow \quad III \quad + 4\,R_2NH \qquad (IV\text{-}1)$$

R = CH₃

III

Several other synthetic procedures have since been developed. Nöth and Regnet[4] reacted phenyldichloroborane, $C_6H_5BCl_2$, with dilithio diphenylhydrazine to yield the tetrazadiborine ring system.

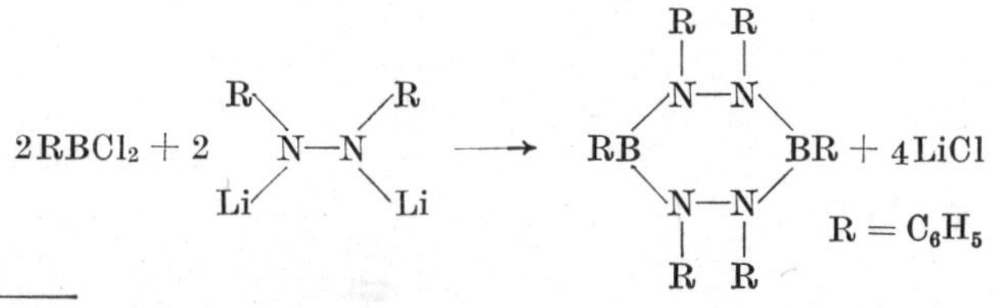

$$2\,RBCl_2 + 2 \quad \begin{matrix} R \diagdown \\ \\ Li \diagup \end{matrix} N{-}N \begin{matrix} \diagup R \\ \\ \diagdown Li \end{matrix} \longrightarrow \quad RB \quad BR + 4\,LiCl \qquad (IV\text{-}2)$$

R = C₆H₅

[1] Nöth, H., and W. Regnet: Z. Naturforsch. 18b, 1138 (1963).
[2] Nöth, H.: ibid. 16b, 470 (1961).
[3] Niedenzu, K., H. Beyer and J. W. Dawson: Inorg. Chem. 1, 738 (1962).
[4] Nöth, H., and W. Regnet: Advances in Chemistry 42, 166 (1964).

In an analogous manner, phenyldichloroborane may be reacted with diphenylhydrazine in the presence of triethylamine[1]. The most elegant approach appears to be by the reduction of azobenzene with diborane:

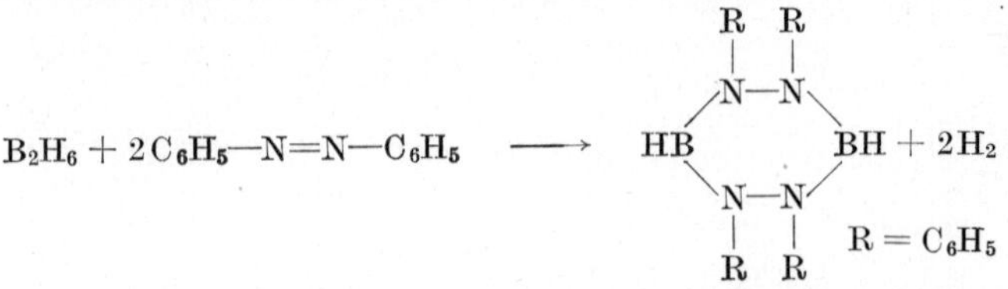

$$B_2H_6 + 2 C_6H_5\!-\!N\!=\!N\!-\!C_6H_5 \longrightarrow \text{(ring structure)} + 2 H_2 \quad (IV\text{-}3)$$

The chemistry of this type of compound has not yet been developed. However, the reaction of bis(dimethylamino)phenylborane with even a large excess of hydrazine does not yield bis(hydrazino)phenylborane[2]. This would appear to confirm the six-membered ring structure as the preferred reaction product. B—N Absorptions in the 1400 cm^{-1} region of the infrared spectra are indicative of a boron-nitrogen bond order >1. Controlled addition of hydrogen chloride to 1,2,4,5-tetraza-3,6-diborines leads to 2:1 and 4:1 adducts[3]. These adducts have been formulated as cyclic structures, IV and V. At higher temperatures, an excess of hydrogen chloride results in the cleavage of the boron-nitrogen bonds.

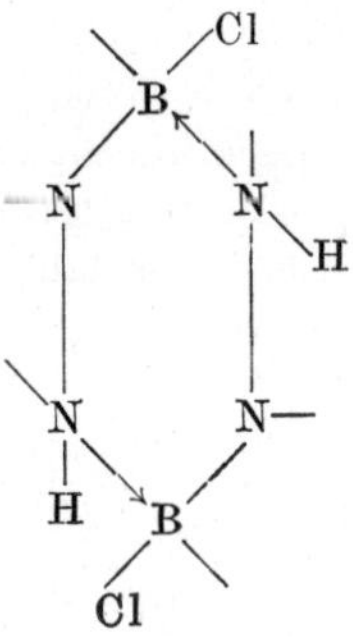

IV 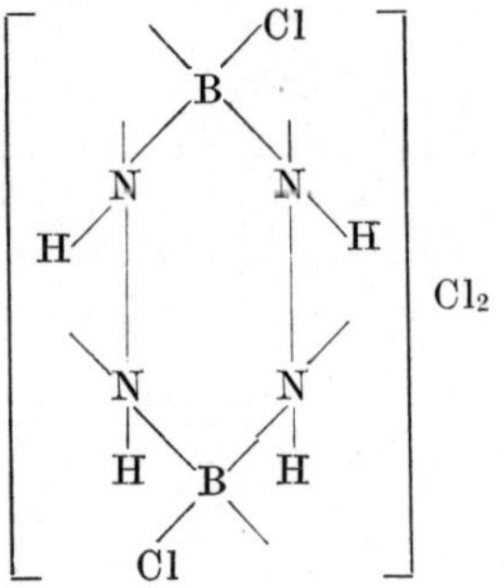   V

E. s-Tetraza-tetraborines

The high bond energy of the normal covalent B—N linkage (104.3 kcal./mole) is in consonance with the relatively great thermal stability of certain boron-nitrogen compounds. One drawback for the application of many boron-nitrogen compounds as high temperature resistant materials resides in their hydrolytic instability. In their search

[1] LAPPERT, M. F., M. K. MAJUMDAR and B. PROKAI: International Symposium on "Boron-Nitrogen Chemistry", Durham, N.C., USA., 1963.

[2] NIEDENZU, K., H. BEYER and J. W. DAWSON: Inorg. Chem. **1**, 738 (1962).

[3] NÖTH, H., and W. REGNET: Z. Naturforsch. **18 b**, 1138 (1963).

for derivatives of greater hydrolytic stability, TURNER and coworkers[1] discovered the s-tetraza-tetraborine system. Dehydrohalogenation of primary amine-trichloroboranes normally results in the formation of borazines (see Chapter III) via the intermediate aminodichloroboranes (see Chapter II-B). However, if highly hindered primary amines are used in these reactions, especially those where the α-carbon atom is fully substituted (i.e. a primary t-alkylamine), the borazine system is not obtained. Linear products may be formed, but, depending on the nature of the substituents, tetrameric azaborines, $(-BCl-NR-)_4$, may result. From the comparatively limited series of examples known so far, it appears that branching at the β-carbon atom of the amine inhibits the formation of the s-tetraza-tetraborines (VI).

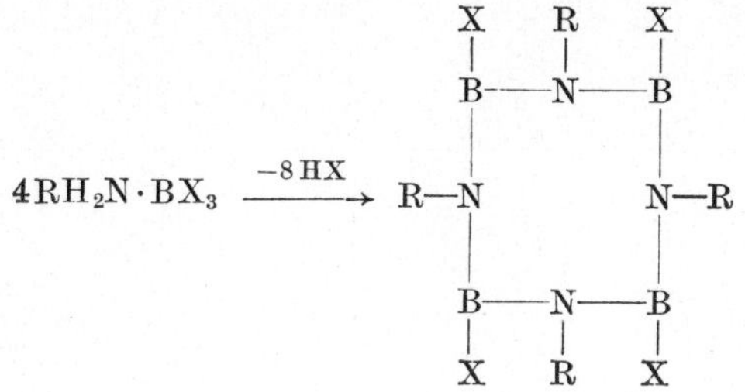

$$
4\,RH_2N \cdot BX_3 \xrightarrow{\ -8\,HX\ }
\begin{array}{ccccc}
X & & R & & X \\
| & & | & & | \\
B & \!\!-\!\!-\!\! & N & \!\!-\!\!-\!\! & B \\
| & & & & | \\
R\!\!-\!\!N & & & & N\!\!-\!\!R \\
| & & & & | \\
B & \!\!-\!\!-\!\! & N & \!\!-\!\!-\!\! & B \\
| & & | & & | \\
X & & R & & X \\
\end{array}
\qquad (IV\text{-}4)
$$

VI

A number of possible structures may be written for a material of the empirical formula $B_4X_4N_4R_4$. However, all four boron atoms in s-tetraza-tetraborines are equivalent only in structure VI and a possible cubane alternate VII. The equivalence of the boron atoms was demonstrated

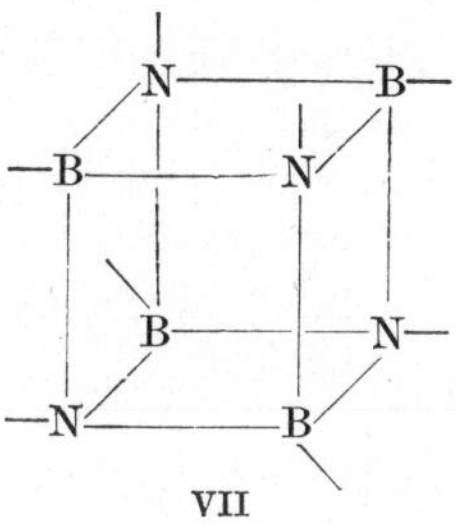

VII

by [11]B nuclear magnetic resonance spectroscopy[2]. Since a molecular model can be made of VI but not of VII, the former structure appears more probable. Dipole moments of s-tetraza-tetraborines were found to be zero or at least extremely small. This datum again is in agreement with either of the above structures. However, X-ray studies[3] substantiate

[1] TURNER, H. S., and R. J. WARNE: Proc. chem. Soc. (London) **1962**, 69.
[2] TURNER, H. S., and R. J. WARNE: Advances in Chemistry **42**, 290 (1964).
[3] CLARKE, P. T., and H. M. POWELL: Abstract of Papers, XIXth International Congress of Pure and Applied Chemistry, London, 1963, p. 202.

the eight-membered ring structure, which probably exists in the boat form with highly localised alternating π-bonds[1].

Many aspects of the chemistry of s-tetraza-tetraborine remain to be elucidated. However, compounds of this type are remarkably stable towards chemical attack. The most thoroughly studied compound, $[-BCl-NC_4H_9(t)-]_4$, is not measurably attacked by boiling water. Attempts to replace chlorine by organic groups via metalorganic reagents have failed and it is assumed that the low reactivity is due to steric hindrance[2]. Some pseudohalogens have been found capable of replacing chlorine and to yield well characterized compounds. Thus the tetra-B-isocyanato derivative has been prepared by the reaction of $[-BCl-NC_4H_9(t)-]_4$ with an alkali-metal isocyanate in a convenient organic solvent.

Chapter V

Heterocyclic σ-Bonded Systems Containing Boron and Nitrogen

A. Introductory

A new area of boron-nitrogen chemistry has developed within the last decade. The basis of this area of investigation, involving B—N—C heterocyclic systems, has been characterized through studies by GOUBEAU and coworkers[3, 4]. In 1955 they prepared 1,3,2-diazaboracycloalkanes illustrating that it is possible to incorporate boron and nitrogen adjacent to each other in a B—N—C ring system in which all annular atoms are σ-bonded to each other[3]. The primary aim of this work was to elucidate the resonance relations in cyclic aminoborane systems. However, the boron analog of indole, 2-methyl-1,3,2-benzodiazaborolidine, which was prepared in 1957[4], can be considered the first known heteroaromatic boron-nitrogen compound. In 1958 DEWAR initiated studies of heteroaromatic compounds in which an aromatic C—C grouping is replaced by the isoelectronic B—N combination. In the following presentation, heteroaromatics will be limited to the six-membered cyclic derivatives.

---

[1] GREEN, J. H. S., W. KYNASTON and H. M. PAISLEY: Advances in Chemistry **42**, 301 (1964).

[2] TURNER, H. S., and R. J. WARNE: ibid. **42**, 290 (1964).

[3] GOUBEAU, J., and A. ZAPPEL: Z. anorg. allg. Chem. **279**, 38 (1955).

[4] ULMSCHNEIDER, D., and J. GOUBEAU: Chem. Ber. **90**, 2733.

Due to the considerable amount of information available on 1,3,2-benzodiazaboracycloalkanes and related compounds, a discussion of these materials separate from the boron-nitrogen-carbon analogs of benzene and its higher homologs seems to be justified. This is especially true since distinct differences in chemical stability exist between the systems. However, no rigid classification of this area has been attempted.

Cyclic B—N—C compounds of the types indicated above, virtually unknown less than ten years ago, have aroused special interest for their potential in cancer therapy[1], but application has also been proposed for their use as antiseptics, insecticides and fungicides[2, 3].

B. Heterocycles with Aminoborane Linkages

On reacting a linear aliphatic α, ω-diamine such as ethylenediamine with trimethylborane, one obtains amine-boranes of type (I)[4].

$$(CH_3)_3B \cdot H_2N—(CH_2)_n—NH_2 \cdot B(CH_3)_3, \; n = 2$$
$$I$$

Heating the product in a sealed tube to about 250°, results in the elimination of methane and the isolation of an aminoborane II in approximately 30% yield.

$$(CH_3)_2B—NH—(CH_2)_2—NH—B(CH_3)_2$$
$$II$$

This compound polymerizes readily; it can also be obtained through direct heating of a stoichiometric mixture of the amine and borane. Utilizing a 1:1 molar ratio of the starting materials, the compound III is obtained. The latter appears to be more stable than II which might indicate the presence of a cyclic structure for the molecule, stabilized by internal back-coordination as illustrated in IV. The validity of this formulation is substantiated by the facile addition of only one mole

$$(CH_3)_2B—NH—(CH_2)_2—NH_2$$
$$III$$

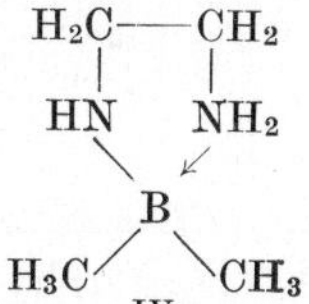

IV

of hydrogen chloride (attacking the aminoborane linkage); on heating IV to 475°, methane is eliminated and the all-σ-bonded 1,3,2 diazaboracycloalkane V is obtained.

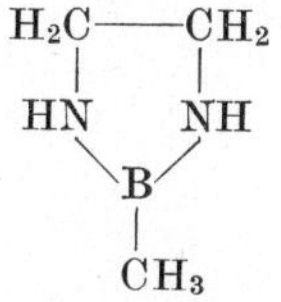

V

[1] NYILAS, E., and A. H. SOLOWAY: J. Amer. chem. Soc. 81, 2681 (1959).
[2] CONKLIN, G. W., and R. C. MORRIS: Brit. Patent 790,090 (1958).
[3] GARNER, P. J.: U.S. Patent 2,839,564 (1958).
[4] GOUBEAU, J., and A. ZAPPEL: Z. anorg. allg. Chem. 279, 38 (1955).

More conveniently, 1,3,2-diazaboracycloalkanes are synthesized by the transamination of a bisaminoborane with the proper aliphatic diamine[1, 2, 3] (eq. V-1).

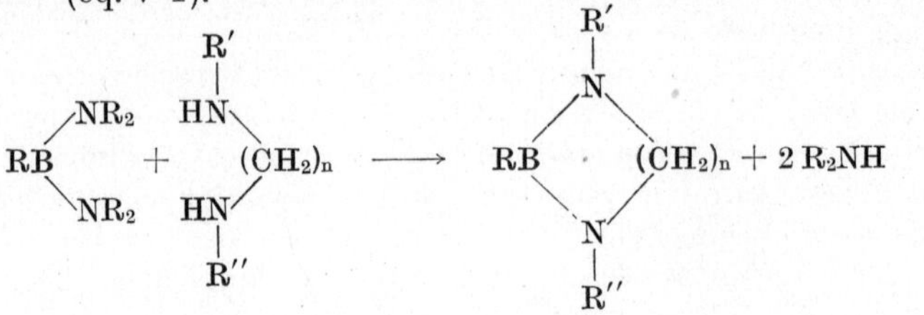

$$
\text{RB}\begin{array}{c}\text{NR}_2\\[2pt]\text{NR}_2\end{array} \;+\; \begin{array}{c}\text{HN}\\[2pt]\text{HN}\end{array}\!\!\!\begin{array}{c}\text{R}'\\[2pt](\text{CH}_2)_n\\[2pt]\text{R}''\end{array} \longrightarrow \quad \text{RB}\;\;\begin{array}{c}\text{R}'\\ \text{N}\\ (\text{CH}_2)_n\\ \text{N}\\ \text{R}''\end{array} + 2\,\text{R}_2\text{NH} \qquad \text{(V-1)}
$$

VI

In recent years, several compounds of the general structure VI have been prepared. These are illustrated in Table V-1.

Table V-1. *1,3,2-Diazaboracycloalkanes (VI)*

R	R'	R''	n	m.p., °C	b.p. (mm.), °C	References
CH_3	H	H	2	43.5	106	1, 2, 5
$CH{=}CH_2$	CH_3	CH_3	2		35—36	4
C_6H_5	H	H	2	157		6
C_6H_5	CH_3	CH_3	2		73 (3)	2
C_6H_5	C_2H_5	C_2H_5	2		95 (8)	2
CH_3	H	H	3		132	5
$CH{=}CH_2$	H	H	3		41 (7)	3
C_2H_5	H	H	3		44—45 (13)	3
C_6H_5	H	H	3	50—52	94—95 (1)	3
C_6H_5	H	H	4	27—29	102—104 (1)	3

However, the transamination reaction is not confined to polymethylene diamines. For example, bisaminoboranes react with 3,3'-diamino-dipropylamine in dilute solution to afford VII.

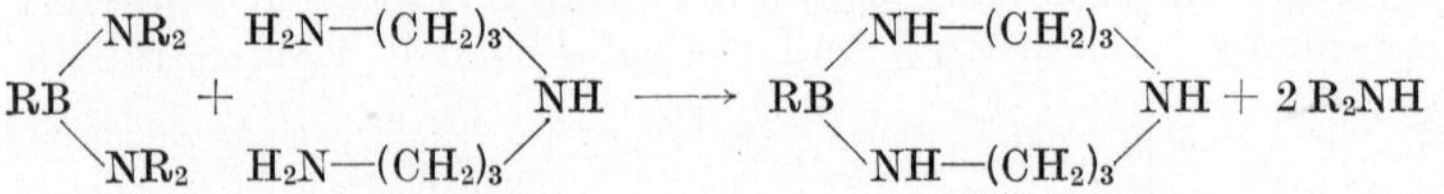

$$
\text{RB}\begin{array}{c}\text{NR}_2\\[2pt]\text{NR}_2\end{array} \;+\; \begin{array}{c}\text{H}_2\text{N}{-}(\text{CH}_2)_3\\ \text{NH}\\ \text{H}_2\text{N}{-}(\text{CH}_2)_3\end{array} \longrightarrow \text{RB}\begin{array}{c}\text{NH}{-}(\text{CH}_2)_3\\ \text{NH}\\ \text{NH}{-}(\text{CH}_2)_3\end{array} + 2\,\text{R}_2\text{NH} \qquad \text{(V-2)}
$$

VII

Another novel ring system, VIII, has been prepared through substi-

[1] Nöth, H.: Z. Naturforsch. **16 b**, 470 (1961).
[2] Niedenzu, K., H. Beyer and J. W. Dawson: Inorg. Chem. **1,** 738 (1962).
[3] Niedenzu, K., P. Fritz and J. W. Dawson: ibid. **3,**, 1077 (1964).
[4] Fritz, P., K. Niedenzu and J. W. Dawson: ibid. **3**, 626 (1964).
[5] Goubeau, J., and A. Zappel: Z. anorg. Chem. **279**, 38 (1955).
[6] Pailer, M., and W. Fenzl: Mh. Chem. **92**, 1294 (1961).

tuting the bisaminoborane with a trisaminoborane[1] in an analogous transamination reaction.

$$\begin{array}{ccc} & CH_2 \quad CH_2 & \\ H_2C & N & CH \\ | & | & | \\ H_2C & B & CH_2 \\ & N \quad N & \\ & H \quad H & \\ & \text{VIII} & \end{array}$$

m.p. $= 38-41°$

b.p. $= 62°(1\ mm)$

The participation of both VII and VIII as reactants in projected studies opens a new area of research.

Two moles of hydrogen chloride add to one of 1,3,2-diazaboracycloalkanes, VI, at the boron-nitrogen bond. Additional treatment of the resultant amine-borane types of compounds with hydrogen chloride at higher temperatures results in cleavage of the B—N bonds[2, 3]. Known derivatives of type VI are thermally quite stable as is indicated by the preparative procedure cited above. In view of the facile preparation of 1,3,2-benzodiazaborolidines from dihydroxyboranes and o-phenylenediamine (see Chapter V-D), it is surprising that 1,3,2-diazaborocycloalkanes are not available by an analogous procedure[4]; only polymeric products were isolated when such a reaction was attempted.

The interpretation of RAMAN spectral data from VI, n = 2,3, confirms the cyclic structure of these materials[2]. The extremely high $\nu_{BN, ring}$ in both derivatives is quite striking (Table V-2). Both values are extraordinarily high as compared to acyclic aminoboranes and to borazines. This departure can be explained in terms of coupling between the boron-attached methyl group with a B—N double bond. In strained ring systems, coupling of methyl groups with double bonds can result in a shift of the double bond absorption to higher frequencies than would normally be expected. In this connection, it is worth noting that it has been shown recently[1] that the bonding between boron and nitrogen in the diazaboracycloalkane system is more like that encountered in borazines than with acyclic bisaminoboranes.

Two 1,2-azaboracycloalkanes (IX), n = 3,4, have recently been described by D. G. WHITE[5]. The compound with n = 3,

Table V-2.

B—N Ring Vibration of 1,3,2-Diazaboracycloalkanes, VI (R'=R''=H)

n	ν (cm.$^{-1}$)
2	1583
3	1488

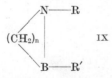

$$\begin{array}{ccc} & N\!\!-\!\!R & \\ (CH_2)_n & & IX \\ & B\!\!-\!\!R' & \end{array}$$

[1] NIEDENZU, K., P. FRITZ and J. W. DAWSON: Inorg. Chem. **3**, 1077 (1964).
[2] GOUBEAU, J., and A. ZAPPEL: Z. anorg. Chem. **279**, 38 (1955).
[3] NIEDENZU, K., H. BEYER and J. W. DAWSON: Inorg. Chem. **1**, 738 (1962).
[4] KATO, S., M. WADA and Y. TSUZUKI: Bull. chem. Soc. Japan **35**, 1124 (1962).
[5] WHITE, D. G.: J. Amer. chem. Soc. **85**, 3634 (1963).

$R = CH_3$, $R' = C_6H_5$, was obtained by (a) the reaction of (allylmethyl-amino)phenylchloroborane with lithium aluminum hydride and (b) by the addition of N-methylallylamine to trimethylamine-phenylborane in diglyme solution. In a procedure analogous to (b), 2-phenyl-1,2-aza-boracyclohexane (n = 4, R = H, $R' = C_6H_5$) was prepared from 3-butenylamine. The remarkable feature of the latter product resides in the

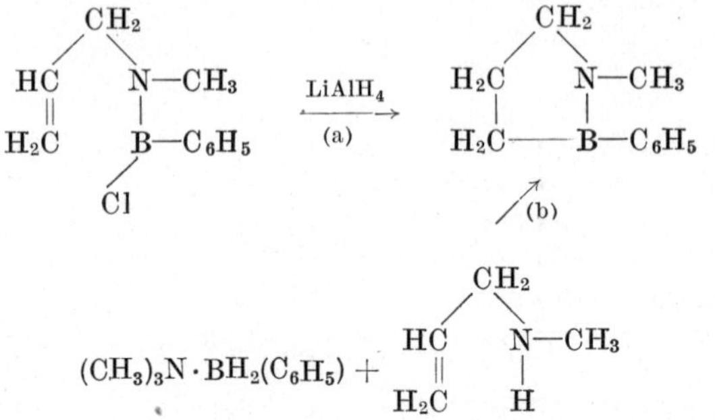

$$(V-3)$$

fact that it can be dehydrogenated to afford 2-phenyl-2,1-borazarene (see Chapter V-D).

Besides the 1,3,2-diazaboracycloalkanes and the 1,2-azaboracyclo-alkanes, a third group of similar compounds has recently been described In the presence of a tert. amine catalyst, aminoboranes of the type RHN—BR$_2'$ react with tetraalkyldiboranes with formation of borazines and the elimination of hydrogen[1].

$$3(R_2BH)_2 + 6R'HN\text{—}BR_2 \xrightarrow{130°} 2(\text{—}BR\text{—}NR'\text{—})_3 + 6BR_3 + 6H_2^* \quad (V-4)$$

In the absence of the tertiary amine, however, formation of a new type of boron-nitrogen-carbon heterocycle, 1,3,2-diboraazacycloalkanes, was effected. For example, in the absence of the amine catalyst, the reaction described in eq. (V-4) ($R = C_6H_5$, $R' = C_2H_5$) does not occur. Rather, a 1,3,2-diboraazacycloalkane of type X is formed.

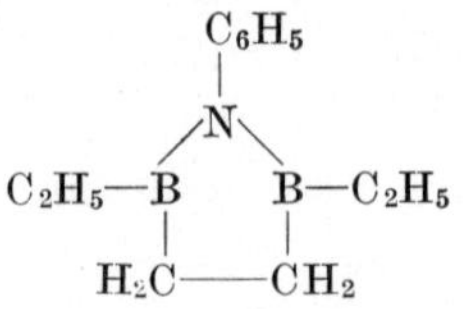

The yields of this new type of heterocycle can be increased to about 40% if the pyrolysis is performed at 200 atm. pressure and in the presence of some trialkylborane. Thus a derivative of 1,3-dibora-2,4-di-

---

[1] Köster, R., and K. Iwasaki: Advances in Chemistry **42**, 148 (1964).

azanaphthalene, XI, was obtained from a mixture of ethyldiborane, triethyldiborane and (phenylamino)diethylborane.

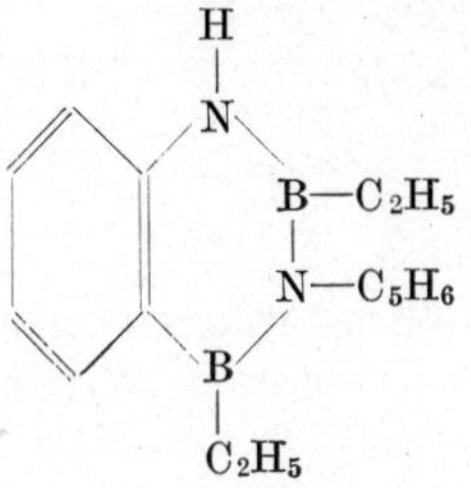

H
|
N
$B-C_2H_5$ m.p. $= 62-63.5°$
$N-C_5H_6$ b.p. $= 143°$ (0.2 mm.)
B
|
C_2H_5
XI

This type of ring formation reaction can also be extended to other alkylboranes and aminoalkylboranes, and the alkylboranes can be replaced by certain boron hydrides. This variation is illustrated by the pyrolysis of a mixture of N-methylbenzylamine and triethylamine-borane. At 200°, hydrogen and triethylamine were evolved and subsequent rectification of the reaction product yielded the following major products (XII, XIII, XIV).

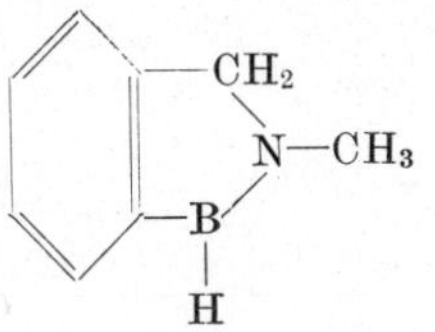

CH_2
$N-CH_3$
B
H
XII

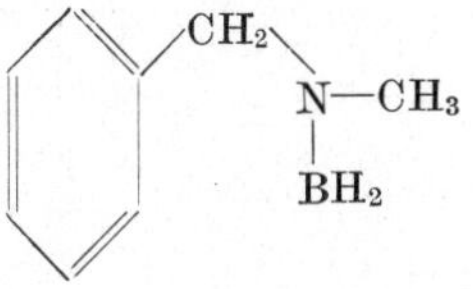

CH_2
$N-CH_3$
BH_2
XIII

$HB[NCH_3-CH_2-C_6H_5]_2$
XIV

A minor amount of XV was also formed. This novel synthesis of a σ-bonded heterocyclic B—N—C system appears to offer extremely interesting prospects for future work.

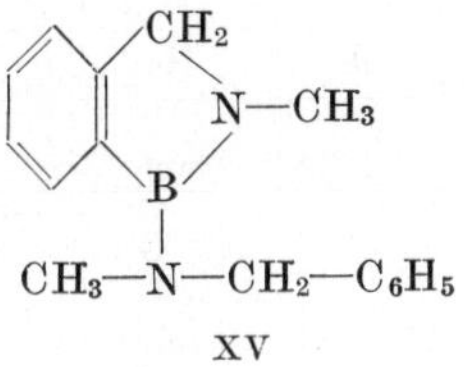

CH_2
$N-CH_3$
B
$CH_3-N-CH_2-C_6H_5$
XV

C. 1,3,2-Benzodiazaborolidines and Related Compounds

Aromatic o-diamines react with trimethylborane in a manner analogous to that described above for aliphatic diamines. Using this procedure, ULMSCHNEIDER and GOUBEAU[1] prepared 1-methyl-1,3,2-

---

[1] ULMSCHNEIDER, D., and J. GOUBEAU: Chem. Ber. **90**, 2733 (1957).

benzodiazaborolidine, XVI, from trimethylborane and o-phenylene-diamine as the first pseudoaromatic B—N—C ring system.

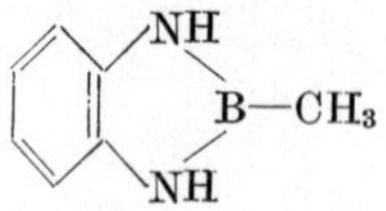

XVI

HAWTHORNE[1, 2] described the reduction of B-trialkylboroxines with lithium aluminum hydride in an ether solution of trimethylamine. The resulting trimethylamine-alkylboranes, $(CH_3)_3N \cdot BRH_2$, are readily converted to the corresponding 2-alkyl-1,3,2-benzodiazaborolidines by treatment with o-phenylenediamine in benzene solution at reflux temperature.

Besides o-phenylenediamine, other o-substituted anilines with reactive hydrogen on the o-substituents, such as o-aminophenol, can be treated in like manner, and their reaction with dihydroxy-, dialkoxy, diamino- and dihalogeno-boranes has been studied in several laboratories[3-12]. This reaction provides ready access to a variety of 1,3,2-benzodiheteroborolidines as illustrated in equation V-5.

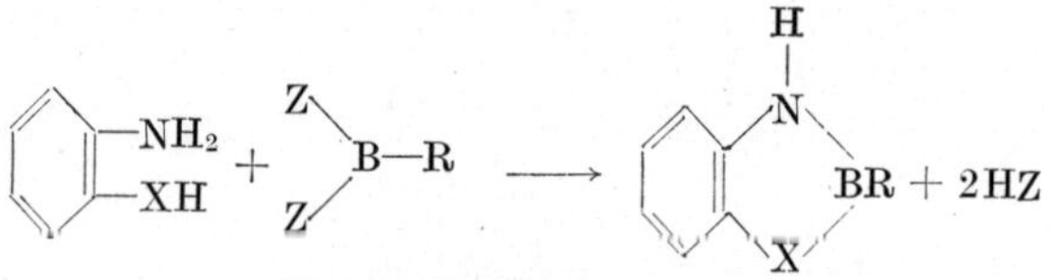

$$X = O, S, NH$$
$$Z = OH, OR, NR_2, \text{halogen}$$

Also, sodium tetraphenylborate has served as a convenient source of borane[13] in this same reaction. A representative listing of 1,3,2-benzo-diheteroborolidines is given in Table V-3.

[1] HAWTHORNE, M. F.: J. Amer. chem. Soc. **81**, 5836 (1959).

[2] HAWTHORNE, M. F.: ibid. **83**, 831 (1961).

[3] LETSINGER, R. L., and S. B. HAMILTON: ibid. **80**, 5411 (1958).

[4] WEIDMANN, H., and H. K. ZIMMERMAN: Liebigs Ann. Chem. **619**, 28 (1958).

[5] DEWAR, M. J. S., V. P. KUBBA and R. PETTIT: J. chem. Soc. (London) **1958**, 3076.

[6] LETSINGER, R. L., and S. H. DANDEGAONKER: J. Amer. chem. Soc. **81**, 498 (1959).

[7] NYILAS, E., and A. H. SOLOWAY: ibid. **81**, 2681 (1959).

[8] CHISSICK, S. S., M. J. S. DEWAR and P. M. MAITLIS: ibid. **81**, 6329 (1959).

[9] HAWKINS, R. T., W. J. LENNARZ and H. R. SNYDER: ibid. **82**, 3053 (1960).

[10] HAWKINS, R. T., and H. R. SNYDER: ibid. **82**, 3863 (1960).

[11] BROTHERTON, R. J., and H. STEINBERG: J. org. Chemistry **26**, 4632 (1961).

[12] BEYER, H., K. NIEDENZU and J. W. DAWSON: ibid. **27**, 4701 (1962).

[13] NEU, R.: Tetrahedron Letters **20**, 917 (1962).

Table V-3. 1,3,2-*Benzodiheteroborolidines*

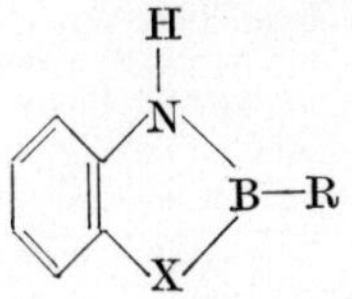

X	R	m.p., °C	References
NH	H	79—80	6
NH	CH_3	98—99	6, 12
NH	$i\text{-}C_3H_7$	124—126	5
NH	C_6H_5	215—216	1–3, 6–10
NH	$CH_2\text{—}C_6H_5$	54—56	4
O	CH_3	32—34	12
O	C_6H_5	105—106	3, 11
S	C_6H_5	154—156	3

The interaction of trichloroborane and o-phenylenediamine was first studied by SCHUPP and BROWN[13], but apparently the authors failed to recognize the true nature of their product. HOHNSTEDT and PELLICCIOTTO[14] demonstrated that the first ring system formed in this reaction is the 2-chloro-1,3,2-benzodiazaborolidine (XVI, R = Cl); this product is relatively unstable and decomposes with the elimination of HCl to yield a borazine derivative XVIII[15, 16]. It has been demonstrated[1]

---

[1] BEYER, H., K. NIEDENZU and J. W. DAWSON: J. org. Chemistry 27, 4701 (1962).

[2] BROTHERTON, R. J., and H. STEINBERG: ibid. 26, 4632 (1961).

[3] DEWAR, M. J. S., V. P. KUBBA and R. PETTIT: J. chem. Soc. (London) 1958, 3076.

[4] HAWKINS, R. T., and H. R. SNYDER: J. Amer. chem. Soc. 80, 3863 (1960).

[5] HAWTHORNE, M. F.: ibid. 82, 748 (1960).

[6] HOHNSTEDT, L. F., and A. M. PELLICCIOTTO: Abstr. of Papers, 137th National Meeting of the American Chemical Society, Cleveland, 1960, p. 7—0.

[7] LETSINGER, R. L., and S. B. HAMILTON: J. Amer. chem. Soc. 80, 5411 (1958).

[8] NEU, R.: Tetrahedron Letters 20, 917 (1962).

[9] NYILAS, E., and A. H. SOLOWAY: J. Amer. chem. Soc. 81, 2681 (1959).

[10] SUGIHARA, J. M., and C. M. BOWMAN: ibid. 80, 2443 (1958).

[11] VAN TAMELEN, E. E., G. BRIEGER and K. G. UNTCH: Tetrahedron Letters 8, 14 (1960).

[12] ULMSCHNEIDER, D., and J. GOUBEAU: Chem. Ber. 90, 2733 (1957).

[13] SCHUPP, L. J., and C. A. BROWN: Abstr. of Papers, 128th National Meeting of the American Chemical Society, Minneapolis, 1955, p. 48-R.

[14] HOHNSTEDT, L. F., and A. M. PELLICCIOTTO: Final Report, U.S. Office of Naval Research, Contract Nonr 2793(00), 1961.

[15] RUDNER, B., and J. J. HARRIS: Abstract of Papers, 138th National Meeting of the American Chemical Society, New York, 1960, p. 61-P.

[16] BROWN, C. A.: Final Report, U.S. Office of Naval Research, Contract Nonr 1939(02), 1956.

that compounds of type XVII readily eliminate RH and undergo condensation (eq. V-6) if R = halogen, OR, or NR_2.

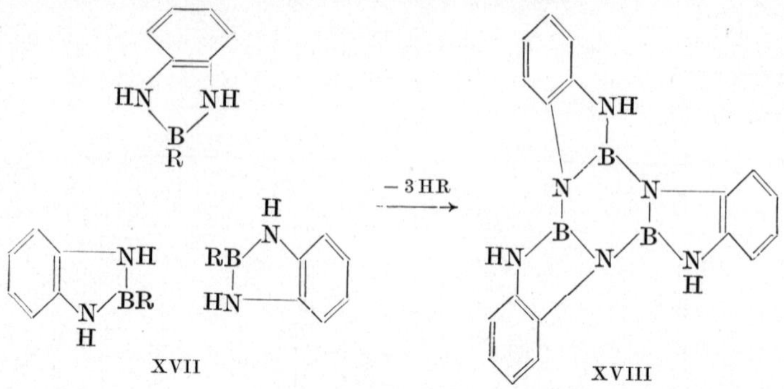

$$(\text{V-6})$$

XVII   XVIII

On this basis, it is understandable that 1,3,2-benzodiazaborolidines of type XVII with R other than alkyl or aryl are not common. The intermolecular condensation reaction to yield the borazine derivative XVIII appears to predominate over normal substitution reactions at the boron atom, thus indicating a preference for this particular ring system. Up to this time only one derivative of XVII with R = NR_2 has been described[1]. It was obtained from tris(dimethylamino)borane and o-phenylenediamine in a transamination reactions in refluxing ether. Refluxing the reactants in a higher boiling solvent caused quantitative formation of XVIII.

A variety of functional aromatic nuclei have been used in place of o-phenylenediamine in analogous reactions as illustrated in eq. (V-5)[2, 3] to yield compounds such as XIX and XX.

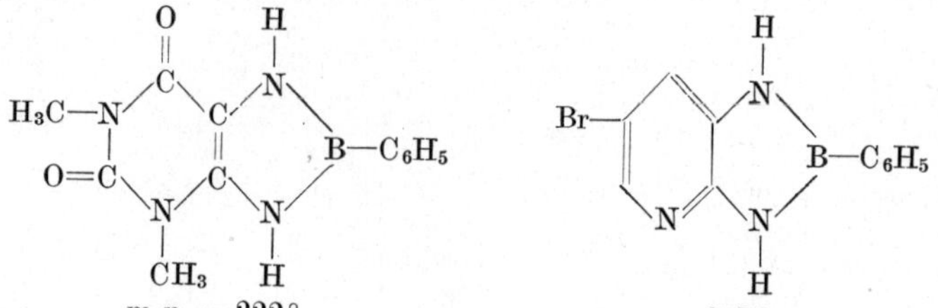

m.p. = 222°
XIX

m.p. = 257°
b.p. = 180° (0.01 mm.)
XX

Some polymers based on the 1,3,2-benzodiazaborolidine system have been reported[4]. In these materials, the basic molecules are coupled by hydrocarbon bridges between two boron atoms and by direct bonding

[1] BEYER, H., K. NIEDENZU and J. W. DAWSON: J. org. Chemistry 27, 4701, (1962).

[2] ZIMMER, H., A. D. SILL and E. R. ANDREWS: Naturwissenschaften 47, 378 (1960).

[3] PAILER, M., and W. FENZL: Mh. Chem. 92, 1294 (1961).

[4] MULVANEY, J. E., J. J. BLOOMFIELD and C. S. MARVEL: J. polymer. Sci. 62, 59 (1962).

of the phenyl rings of adjacent molecules, XXI. They do not seem to have any technical importance since 1,3,2-benzodiheteroborolidines are

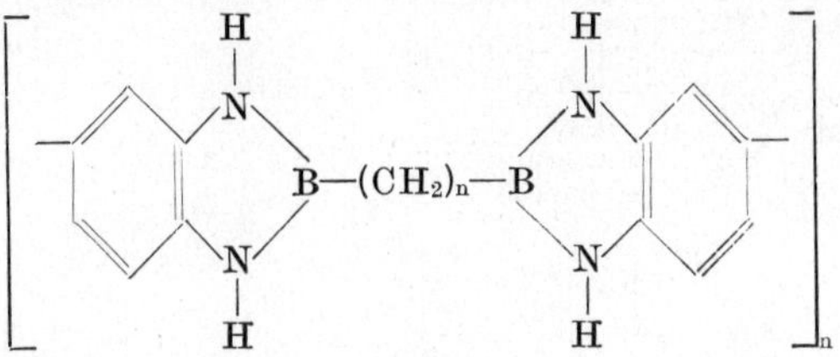

XXI

not stable. They are all readily solvolyzed and undergo autoxidation at the annular boron bonds. The similarity of their properties to those of cyclic amides indicates that there is little aromatic stabilization inherent in the system, although they might have some aromatic character. This is evidenced by similarities of their ultraviolet spectra to those of isoconjugate nonboron-containing heterocycles[1].

It seems proper to cite one cyclic compound in which an olefinic carbon-carbon double bond is conjugated with a boron-nitrogen linkage. LETSINGER and HAMILTON[2] described the interaction of (dihydroxy)-phenylborane with an equimolar mixture of benzoin and aniline which resulted in the formation of 2,3,4,5-tetraphenyl-1,3,2-oxazaborole, XXII.

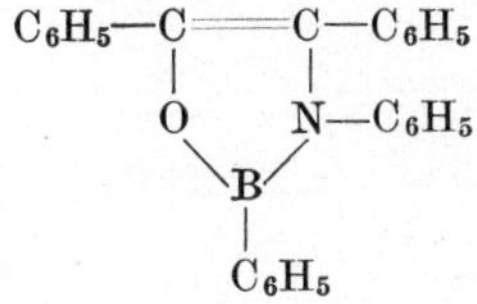

XXII

Six-membered pseudoaromatic ring systems as exemplified by 2-phenyl-2-boradihydroperimidine, XXIII, have also been reported. On comparing the ultraviolet spectrum of this compound with that of 2-phenylperi-midine[3], XXIII does not evidence the existence of aromatic character to any appreciable degree.

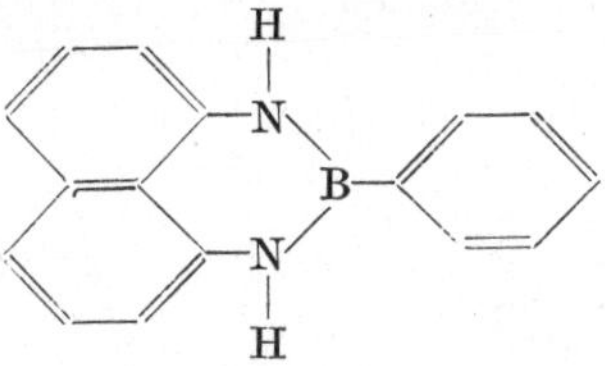

XXIII

[1] DEWAR, M. J. S., V. P. KUBBA and R. PETTIT: J. chem. Soc. (London) 1958 3073.

[2] LETSINGER, R. L., and S. B. HAMILTON: J. org. Chemistry 25, 592 (1960).

[3] CHISSICK, S. S., M. J. S. DEWAR and P. M. MAITLIS: Tetrahedron Letters 23, 8 (1960).

One last compound deserves mention; this is the one obtained through the SOMMELETT reaction on α-bromo-2-tolyl-dihydroxyborane and to which the following structure XXIV has been assigned[1].

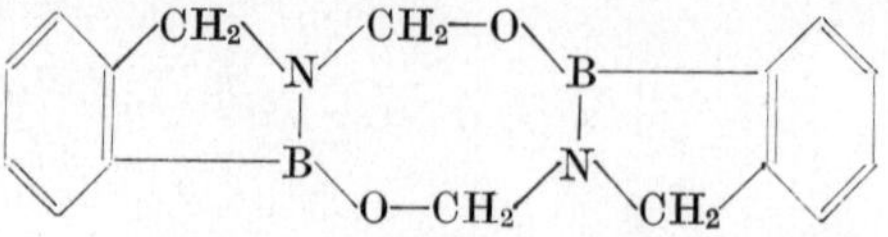

XXIV

D. Heteroaromatic Boron-Nitrogen Compounds

1. General Remarks

Theoretically, heteroaromatic boron-nitrogen compounds can be derived from any regular aromatic compound by replacement of two carbon atoms by one boron and one nitrogen atom. For example, from benzene one can formulate 2,1-borazarene, XXV, which, like the isoelectronic benzene, should be aromatic and resonance-stabilized. Two additional structures of identical analytical composition can be formulated, XXVI, XXVII; they differ from XXV in that the boron and nitrogen atoms in the ring are separated by one or two carbon atoms.

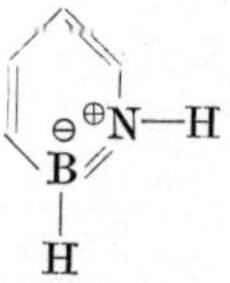

XXV 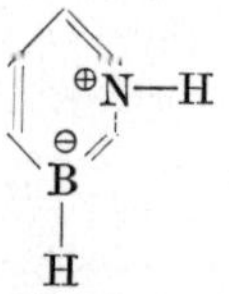   XXVI 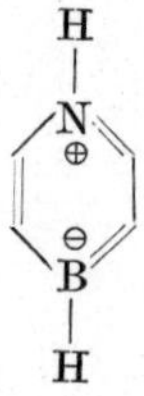   XXVII

Of these three isomers, XXV and XXVII should be the most stable, since uncharged resonance structures can be formulated for them (XXVIII, XXIX). Also, LCAO molecular orbital calculations confirm an order of decreasing thermochemical stability XXV > XXVII > XXVI[2]. Indeed, derivatives of XXV and XXVII have been synthesized, but so far no compounds of type XXVI are known; no hetero-

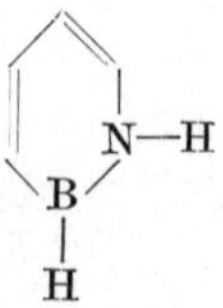

XXVIII 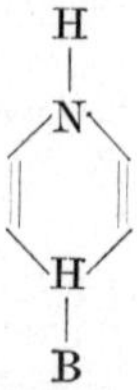   XXIX 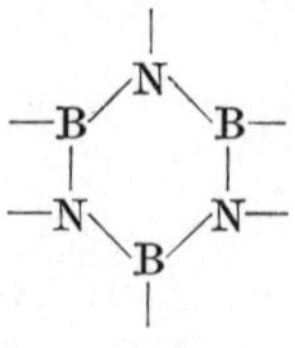   XXX

  [1] SNYDER, H. R., A. J. REEDY and W. J. LENNARZ: J. Amer. chem. Soc. 80, 835 (1958).
  [2] HOFFMANN, R.: Advances in Chemistry 42, 78 (1964).

aromatic compounds with two boron and two nitrogen atoms in a benzene ring have yet been reported, whereas, borazines, XXX, have been known for more than thirty years.

The notable feature of the heteroaromatic B—N—C system as compared to the non-aromatic B—N—C heterocycles resides in the unusual chemical stability of the former. This feature, in conjunction with other properties, makes heteroaromatics a distinct and interesting group of new compounds. For the sake of convenience and consideration, they are categorized in relation to their hydrocarbon analogs.

2. 2,1-Borazarenes

The most important heteroaromatic boron-nitrogen compounds which have so far been synthesized are homologs of 2,1-borazarene. The first compounds of this series, derivatives of the 10,9-borazaro-

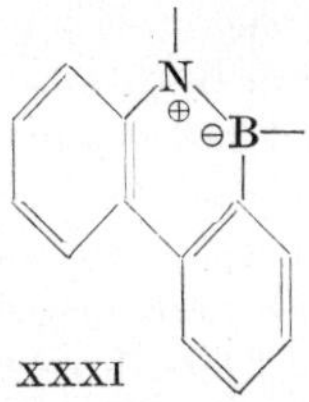

XXXI

phenanthrene, XXXI, were prepared by DEWAR and coworkers in 1958[1]. They were recognized at that time to be much more stable than B—N—C heterocycles of the types described above (e.g. 1,3,2-diheteroborolidines) and linear boron-nitrogen compounds showing a similar type of B—N bonding (aminoboranes); the boron-nitrogen heteroaromatics do not lend themselves to the addition reactions so characteristic of the borazine system (see Chapter III). The greatly enhanced stability of the B—N heteroaromatics has been attributed to aromatic stabilization of the hetero ring[2]. This view is supported by their ultraviolet spectra which are very similar to those of the corresponding aromatic hydrocarbons. In addition, it was found that 2,1-borazarenes readily undergo electrophilic substitution reactions which are considered to be typical of aromatic compounds (e.g. bromination, nitration). The 6 and 8 positions of the 10,9-borazarophenanthrene, XXXI, are the most reactive, but when subjected to more stringent reaction conditions, 2,6,8-trisubstituted products can be prepared. The ring location of the substituents has been shown to be in agreement with molecular orbital calculations[3]. These studies indicate that structures such as XXXI should have only a very small dipole moment, but, at the same time, a large resonance energy. The polarity due to the N→B π-bonding is balanced by the polarization of the π-electrons in the terminal benzene rings.

[1] DEWAR, M. J. S., V. P. KUBBA and R. PETTIT: J. chem. Soc. (London) 1958, 3073.

[2] HOFFMANN, R.: Advances in Chemistry 42, 78 (1964).

[3] DEWAR, M. J. S., and V. P. KUBBA: Tetrahedron 7, 213 (1959).

The main preparative route to 2,1-borazarene homologs involves the reaction of primary aromatic amines with dihalogenoboranes as illustrated in eq. V-7.

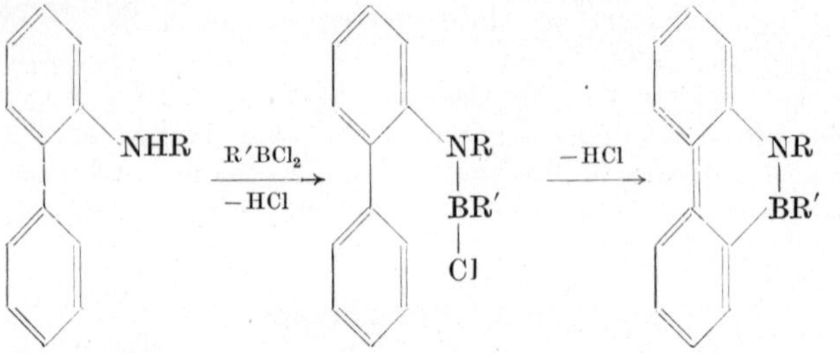

$$\text{(V-7)}$$

A large number of variously substituted 2,1-borazarenes have been prepared by this and closely related methods. It can not be used for the preparation of monocyclic derivatives. DEWAR and MARR[1] reported the successful synthesis of one such monocyclic derivative, but the method involved does not appear to be of general use. More recently[2], a preparation of 2-phenyl-2,1-borazarene by dehydrogenation of the corresponding 1,2-azaboracyclohexane (see Chapter V-B) has been reported and seems to have a high potential for future synthetic studies.

The same preparative route as illustrated in eq. V-7 has also been used to provide access to the 2,1-borazaronaphthalene system, XXXII. The chemical properties and physical characteristics of heteroaromatics of

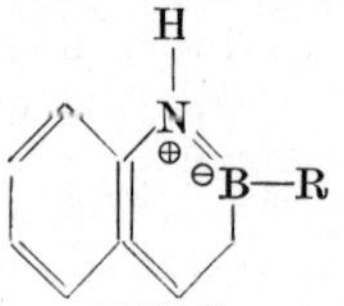

XXXII

Table V-4. *2,1-Borazaronaphthalenes*

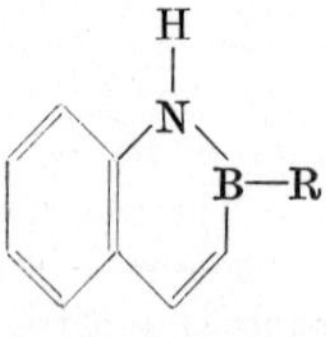

R	m.p., °C	References
H	100—101	3, 4
CH$_3$	73—74	3, 4
C$_6$H$_5$	138—139	3
Cl	72—74	3
OCH$_3$	57—58	4

type XXXII closely parallel those of the isoelectronic naphthalene derivatives. However, 2,1-borazaronaphthalenes are less plentiful than the 10,9-borazarophenanthrenes. This situation appears to be due to difficulties in preparing these compounds and is not related to their stability. The parent compound (XXXII, R = H) is stable to acids, in contrast to the 10,9-borazarophenanthrene[3]. Table V-4 includes derivatives of type XXXI.

[1] DEWAR, M. J. S., and P. A. MARR: J. Amer. chem. Soc. 84, 3782 (1962).
[2] WHITE, D. G.: ibid. 85, 3634 (1963).
[3] DEWAR, M. J. S., and R. DIETZ: J. chem. Soc. (London) 1959, 2728.
[4] DEWAR, M. J. S., R. DIETZ, V. P. KUBBA and A. R. LEPLEY: J. Amer. chem. Soc. 83, 1754 (1961).

The synthetic procedure described above is not limited to the availability of aromatic starting materials having only annular carbon atoms; moreover, heteroaromatics are known with more than one boron-nitrogen grouping in the molecule. For instance, 4,10-dibora-5,9-diazaropyrenes. **XXXIII**, and 2,7-dibora-1,8-diazaroanthracenes, **XXXIV**, have been prepared[1, 2] and their spectra recorded.

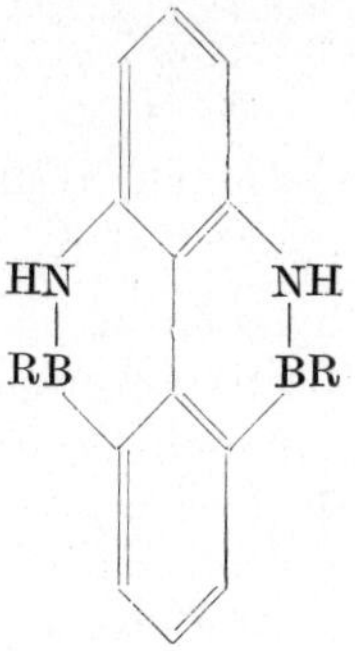

XXXIII 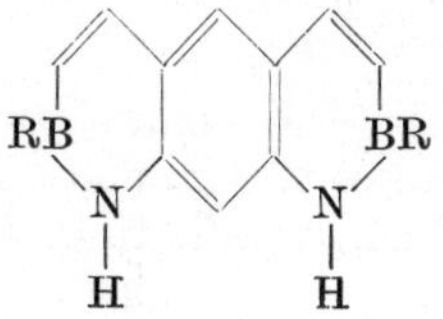   XXXIV

3. 4,1-Borazarenes

4,1-Borazarenes are known to exist only as higher homologs, i.e. 10,9-borazaroanthracenes (**XXXV**). They can be synthesized by the

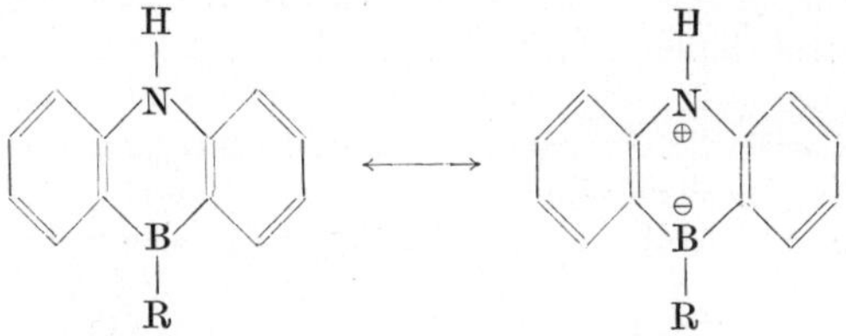

XXXV(a)   XXXV(b)

action of boroxines, $(—BR—O—)_3$, upon 2,2′-dilithiodiphenylamine. These compounds, containing isolated boron and nitrogen atoms, are less

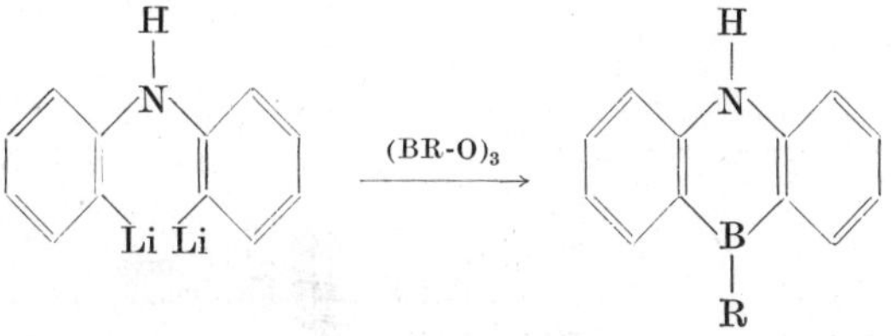

(V-8)

---

[1] CHISSICK, S. S., M. J. S. DEWAR and P. M. MAITLIS: Tetrahedron Letters **23,** 8 (1960).

[2] MAITLIS, P. M.: Chem. Review **62,** 223 (1962).

stable than the 2,1-borazarenes discussed above. This characteristic suggests that the incorporation of a boron and a nitrogen atom into an aromatic ring leads to the formation of a stable hetero system only where the two heteroatoms are adjacent and is in agreement with molecular orbital calculations[1-3]. It is possible, therefore, that the uncharged classical formulation, as illustrated in XXXVa, is predominant over the truly aromatic system with double bond conjugation.

4. Other Heteroaromatic Boron-Nitrogen Systems

The first heteroaromatic compound with boron and/or nitrogen at the bridgehead of two aromatic nuclei was 12,11-borazarophenanthrene, XXXVI[4], which appears to be an example of a stable aromatic system.

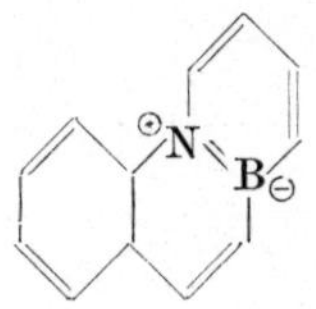

XXXVI 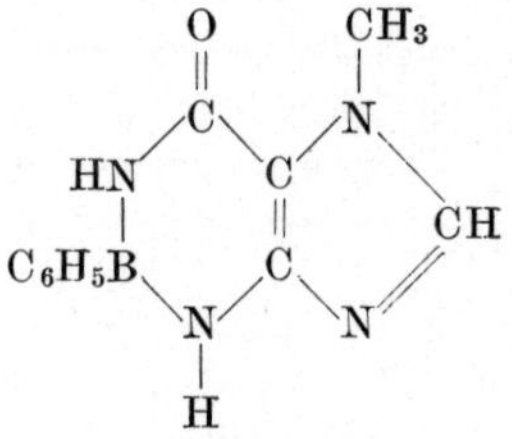   XXXVII

Previous attempts to dehydrogenate 7-methyl-7,16-borazaro-5,6-di-hydrobenz[a]anthracene had failed[3]. However, the parent system, 10,9-borazaronaphthalene has been cited earlier[3].

Recently, boron-nitrogen analogs of quinoline[4] and isoquinoline[5] have become available. It has been found that (dihydroxy)-o-formyl-phenylborane reacts with phenylhydrazine to afford an almost quantitative yield of 3-phenyl-4-hydroxy-4,3-borazaroisoquinoline[5].

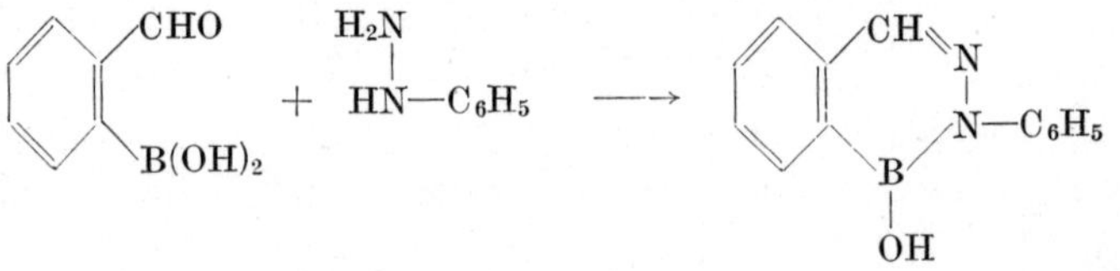

In this connection, a boron-containing purine analog (XXXVII)[6] should be recognized, although it cannot be classed as heteroaromatic.

[1] DEWAR, M. J. S., and R. DIETZ: J. chem. Soc. (London) 1959, 2728.
[2] KAUFMAN, J. J., and J. R. HAMANN: Advances in Chemistry 42, 273 (1964).
[3] DEWAR, M. J. S.: ibid. 42, 227 (1964).
[4] SOLOWAY, A. H.: J. Amer. chem. Soc. 82, 2442 (1960).
[5] DEWAR, M. J. S., and R. C. DOUGHERTY: ibid. 84, 2648 (1962).
[6] CHISSICK, S. S., M. J. S. DEWAR and P. M. MAITLIS: ibid. 81, 6329 (1959).

The above resume illustrates the wide preparative possibilities in this area of boron-nitrogen chemistry. However, these new compounds are also extremely interesting from a theoretical point of view. Their π-electron distributions are strongly perturbed in comparison with the classical aromatic systems. It is obvious that this situation can provide a fertile ground for an examination of chemical theory in future work. Some work in this direction has already been done[1]. Surprisingly, and in contrast to calculations on borazines, molecular orbital calculations of heteroaromatic boron-nitrogen compounds agree very well with the present experimental data if the input parameters are based on a model involving $B^{\ominus}-N^{\oplus}$ species. This difference between borazaromatics and borazines must be due, at least in part, to the differences in the σ-framework of the two systems. It is reasonable to assume that a boron atom between a carbon and a nitrogen atom, as is the case in the heteroaromatic boron-nitrogen compounds, will have an initially higher σ-electron density than a boron atom, which is bonded to two nitrogen atoms. This is based on the smaller electronegativity of carbon as compared to nitrogen, which results in a lessened power of the carbon to attract electrons from the boron. On this same basis, a nitrogen atom in a borazine ring will have a higher σ-electron density than a nitrogen atom in the cyclic structure of borazaromatics. Hence, such changes in the σ-electron distribution within the boron-nitrogen entity may be responsible for the differences in chemical behavior between heteroaromatic boron-nitrogen compounds and borazines.

Chapter VI

·

Boron Nitride

A. Preparation and Structure

Boron nitride, $(BN)_x$, is one of the oldest known boron-nitrogen compounds. Its close resemblance to carbon in its various structural forms combined with certain physical characteristics makes boron nitride a material of considerable theoretical interest and more recently of technological importance.

---

[1] KAUFMAN, J. J., and J. R. HAMANN: Advances in Chemistry **42**, 273 (1964).

The preparation of boron nitride has been studied quite extensively but even the most recent synthetic procedures involve considerable technical difficulties. A number of reports, dating back to the middle of the last century, on the preparation of boron nitride from elemental boron or boron oxide and nitrogen, nitrogen oxides, or ammonia can be found in the literature. Other methods of preparation involve the fusion of boric oxide (or borates) with cyanides, metal amides, or urea, while still others describe the thermal decomposition of trihalogeno-borane ammonolysis products.

STOCK and HOLLE[1] ammonolyzed tribromoborane in liquid ammonia and, after the reaction mixture was allowed to warm to room temperature thus evaporating the excess of ammonia, the residue was slowly heated to 750° in an atmosphere of ammonia to yield boron nitride. On analogous treatment of trichloroborane, MEYER and ZAPPNER[2] obtained boron nitride of 99.4% purity.

$$\begin{cases} BX_3 + 6NH_3 \rightarrow [B(NH_2)_3] + 3NHX_4 \\ x\,[B(NH_2)_3] \rightarrow (BN)x + 2x\,NH_3 \end{cases} \qquad \begin{aligned} &(VI\text{-}1) \\ &(VI\text{-}2) \end{aligned}$$

A laboratory preparation of pure boron nitride, involving the fusion of borax with ammonium chloride, was reported by TIEDE and TOMA-SCHEK[3].

Technical methods of preparation usually involve fusing urea with boric acid in an atmosphere of ammonia. By this method, a boron nitride of substantial purity is obtained at pyrolysis temperatures of 500–950°[4]. The product has a turbostatic structure which is formally analogous to turbostatic carbon. The turbostatic structure can readily be converted into an ordered-layer-lattice hexagonal form through thermal treatment at temperatures below 1800° when traces of boron oxide are present[5].

The hexagonal modification of boron nitride has a bimolecular unit cell with a simple layer structure. Each layer consists of a flat or nearly flat network of B_3N_3 hexagons. There are four possible ways in which these layers may be packed, three of which are simply different arrangements of boron and nitrogen atoms at sites comparable to those occupied by carbon atoms in graphite. One of these structures has been accepted as the correct conformation of hexagonal boron nitride for a long time[6]. However, PEASE[7] has shown the correct arrangement which comprises the fourth possibility as is seen in Fig. VI-1.

The atomic sites can be written:

$$N: (0,\,0,\,0) \quad \text{and} \quad (\tfrac{1}{3},\,\tfrac{2}{3},\,\tfrac{1}{2})$$
$$B: (0,\,0,\,\tfrac{1}{2}) \quad \text{and} \quad (\tfrac{1}{3},\,\tfrac{2}{3},\,0)$$

[1] STOCK, A.. and W. HOLLE: Ber. dtsch. chem. Ges. 41, 2095 (1908).
[2] MEYER, F., and R. ZAPPNER: ibid. 54, 560 (1931).
[3] TIEDE, H., and H. TOMASCHEK: Z. anorg. allg. Chem. 147, 114 (1925).
[4] O'CONNOR, T. E.: J. Amer. chem. Soc. 84, 1753 (1962).
[5] THOMAS JR., J., N. E.WESTON and T. E. O'CONNOR: ibid. 84, 4619 (1962).
[6] BRAGER, A.: Acta physicochem. USSR. 10, 902 (1939).
[7] PEASE, R. S.: Acta crystallogr. (Copenhagen) 5, 356 (1952).

In contrast to the first three structural possibilities cited, the actual structure of hexagonal boron nitride involves relatively small differences between F_o and F_c. An identical structure has not been found for any other substance, although it is formally related to the zincite structure in the same way as the structure of graphite is related to that of diamond.

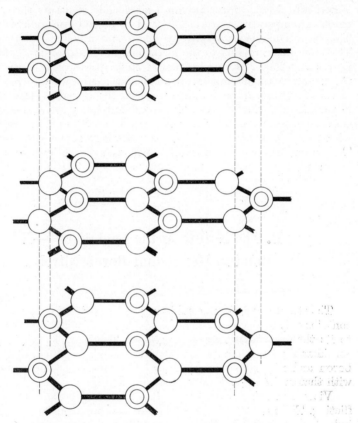

Fig. VI-1: *Layer Arrangement in hexagonal Boron Nitride*

This is of significance since other nitrides of the third group of the Periodic Table (AlN, GaN) also have the zincite structure. Actually, graphite is the only close analog of boron nitride. The geometrical difference between these two materials lies only in the nature of layer packing. In boron nitride, the hexagons are packed directly on top of each other, whereas in graphite, a form of close packing exists in which half the atoms lie between the centers of hexagonal rings of adjacent layers. The similarity between boron nitride and graphite in respect to most properties indicates that the packing difference is of secondary importance; the magnitudes of interatomic forces in the two materials are very similar.

Since the boron-nitrogen bond has an electrical dipole moment, the difference of packing may well be due to the interlayer interaction of the dipoles. Large expansion in the c-direction of boron nitride indicates weak interlayer bonding. The bonding energy between the flat networks is about 4 kcal./mole[1].

The common hexagonal modification of boron nitride can be converted into a cubic form having a zincblende structure through treatment at temperatures near 1800° and at 85,000 atmospheres pressure. This conversion is catalyzed by alkali and alkaline earth metals[2, 3]. The cubic form reverts to the hexagonal modification at 50,000 atmospheres and 2500°. By application of static pressures at lower temperatures, down to room temperature and below, a cubic wurtzite modification of boron nitride can be obtained from the hexagonal modification[4] even in the absence of a catalyst. The disadvantage of this latter structural conversion resides in the fact that only small crystals can be obtained whereas much larger crystals can be realized from the catalytic process. The minimum pressure for any such transformations of hexagonal boron nitride into cubic modifications is about 115 kbar. at 2000° K.

B. The Nature of the Coplanar B—N Bond in Hexagonal Boron Nitride

The structure of the normal hexagonal modification of boron nitride contains only one type of strong bond, the B—N bond joining each atom to its three coplanar nearest neighbors. The bond length is 1.446 Å[5], considerably less than the sum of the single-bond covalent radii of boron and nitrogen (1.58 Å). It is of interest to compare this value with that of 1.44 ± 0.02 Å for borazine[6].

Views as to the nature of the B—N bond in boron nitride are conflicting. The magnitudes of the interatomic forces in boron nitride are indeed very similar to those in graphite, where a partial double-bond model is correct. Consequently, the view of LEVY and BROCKWAY[7] which assumes that the short bond length indicates a partial double bond character, involving resonating electrons similar to the situation in graphite, has many followers. However, this similarity is offset by the

[1] SAMSONOV, G. V., and V. M. SLEPTSOV: Kinetika i Kataliz, Acad. Sci. USSR. Sb. Statei **1960,** 129.
[2] WENTORF, R. H.: J. chem. Physics **26,** 956 (1957).
[3] WENTORF, R. H.: ibid. **34,** 809 (1961).
[4] BUNDY, F. P., and R. H. WENTORF: ibid. **38,** 1144 (1963).
[5] PEASE, R. S.: Acta crystallogr. (Copenhagen) **5,** 356 (1952).
[6] BAUER, S. H.: J. Amer. chem. Soc. **60,** 524 (1938).
[7] LEVY, H., and L. O. BROCKWAY: ibid. **59,** 2085 (1937).

quite different electrical properties of the two materials which are readily interpreted if a singlebonded model is assumed. Thus HÜCKEL[1] has reasoned that the whiteness and low electrical conductivity of boron nitride argue against the presence of resonating electrons and, consequently, only a B—N single bond must be present. It is of interest, that most of the evidence is readily explainable only in terms of a single-bond model; but the actual electronic distribution may well be an intermediate between the two extreme situations.

In most substances, the nature of a bond may be determined unambiguously from bond length data. This procedure is not quite reliable for boron compounds, since the boron atom may be capable of existing without a complete electron octet, in which state its effective radius is smaller. Thus the short bond length as observed for boron nitride can be attributed to the same type of single-bond radius as found in compounds in which the boron atom is surrounded by an electron sextet.

Two structures are therefore proposed for boron nitride, one involving trigonal σ-bonds (I) and the other one based on a double-bond model (II).

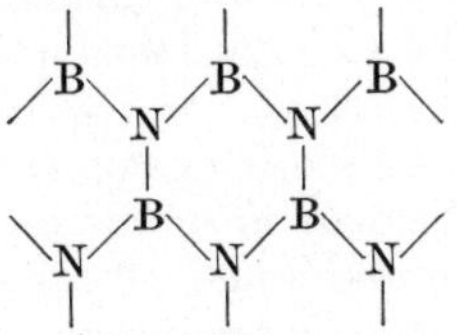

$$
\begin{array}{cc}
\text{I} & \text{II}
\end{array}
$$

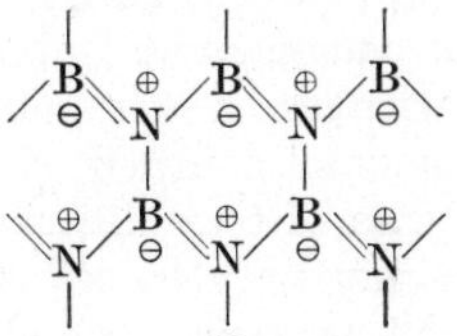

The [11]B nuclear magnetic resonance of boron nitride was observed at 7.177 Mcps.[2]. From second order theory, the magnitude of the quadrupole coupling constant of [11]B in boron nitride is 2.96 ± 0.1 Mcps. Based on the above structures, the quadrupole coupling indicates 55% single bond and 45% double bond configurations, since the possible resonating structure involving a positively charged boron atom, $B^{\ominus}$, bonded to two neutral nitrogens (the third being $N^{\ominus}$) has a coupling constant equal to that of structure I. Consequently, the figure of 45% double bond character in boron nitride with a net charge of -0.45 e of the boron seems to be valid.

BRAGER and SHDANOW[3] have used one-dimensional FOURIER analysis to determine the distribution of electrons in boron nitride. However, this work was based on a method developed by BRILL who strongly criticized the earlier authors for their conclusions[4]. Recently, the electronic properties of boron nitride have been discussed in terms of localized molecular orbital bonds[5]: the definitions of covalency and polarity of the bonds were linked with the net ionicity of the atoms of the crystal.

[1] HÜCKEL, W.: "Anorganische Strukturchemie". Stuttgart: Emke Verlag 1948.
[2] SILVER, A. H., and P. J. BRAY: J. chem. Physics **32**, 288 (1960).
[3] BRAGER, A., and H. SHDANOW: C. R. (Doklady) Acad. Sci. USSR. **28**, 629 (1940).
[4] BRILL, R., C. HERMANN and C. PETERS: Naturwissenschaften **29**, 784 (1941).
[5] COULSEN, C. A., L. B. REDEI and D. STOCKER: Proc. Roy. Soc. (London), Ser. A **270**, 357 (1962).

C. Physical Properties of Hexagonal Boron Nitride

Boron nitride has three most interesting properties, namely refractory, electrical and lubricating. The lubricating properties are readily explained by reference to the structurally related graphite. The general appearance of boron nitride was accurately described by GOLD-SCHMIDT[1]. His figure of 2.255 ± 0.02 for the specific gravity of boron nitride is in good agreement with a more recent measurement by the displacement method giving a value of 2.29 ± 0.03[2]. The theoretical value obtained from unit-cell dimensions of the normal hexagonal modification is calculated to be 2.270. The refractive index of boron nitride was found to be n > 1.74. In marked contrast to the metallic appearance of graphite, however, boron nitride is white and has exceptionally good insulating properties, which, at higher temperatures, are better than those of most refractory oxides[3]. Thus the specific resistance of boron nitride at 2000° was found to be 1900 Ω[4]; at room temperature a value of $>10^{12} \Omega$ has been reported[5]. The diamagnetic susceptibility has been determined at $-0.4 \pm 0.1 \times 10^{-6}$ c.g.s.e.m.u.g.$^{-1}$ [6], which is of course quite different from the extraordinarily large value found for graphite (-10×10^{-6}).

After several controversies over spectroscopic examinations, the band spectrum of boron nitride was obtained by an electrical discharge through helium with traces of trichloroborane and nitrogen. A singlet and a triplet system were attributed to the boron nitride molecule[7]. The infrared spectrum of boron nitride shows two absorption bands, centered at 1372 and 812 cm.$^{-1}$ [8, 9].

In contrast to the normal hexagonal modification of boron nitride, the cubic form is extremely hard, surpassing even diamond.

D. Chemical Reactions of Boron Nitride

Reports on the chemical reactions of boron nitride are often contradictory. For instance, WÖHLER[10] described the attack of water vapor on boron nitride at high temperatures as leading to a cleavage of the molecule with the formation of ammonia and boric acid. STOCK and co-

[1] GOLDSCHMIDT, V. M.: Norsk geol. Tidsskr. **9**, 258 (1926).
[2] PEASE, R. S.: Acta crystallogr. (Copenhagen) **5**, 356 (1952).
[3] PODSZUS, E.: Elektrotechn. Z. **39**, 229 (1918).
[4] PODSZUS, E.: Angew. Chem. **30**, 156 (1917).
[5] RENNER, TH.: Z. anorg. allg. Chem. **298**, 22 (1958).
[6] DAWSON, J. K.: as cited in ref. (2).
[7] DOUGLAS, A. E., and G. HERZBERG: Canad. J. Res. A **18**, 179 (1940).
[8] BRAME, E. G., J. L. MARGRAVE and V. M. MELOCHE: J. inorg. nucl. Chem. **5**, 48 (1957).
[9] MILLER, F. A., and C. H. WILKINS: Analytic. Chem. **24**, 1253 (1952).
[10] WÖHLER, F.: Liebigs Ann. Chem. **74**, 72 (1850).

workers[1, 2] reported slow hydrolysis of boron nitride even at room temperatures, but FRIEDERICH and SITTIG[3] claimed that a boron nitride prepared at high temperatures is not attacked by water. Also, a patent[4] exists wherein it is claimed that boron nitride, after being heated in an atmosphere of boric oxide to temperatures above 2000°, is extremely stable towards chemical attack. It seems reasonable to assume that the method of preparing boron nitride is of the utmost importance in determining its chemical behavior. This assumption is in agreement with a very early report[5] that two modifications of boron nitride of different chemical reactivity exist. However, no systematic studies are yet available with respect to specific structural features and their relationship to chemical reactivity.

One of the more thoroughly studied reactions concerns the fluorination of boron nitride. GLEMSER and HAESELER[6] reported the interaction of boron nitride with hydrogen fluoride to afford a quantitative yield of ammonium fluoroborate:

$$BN + 4HF \rightarrow [NH_4][BF_4] \tag{VI-3}$$

This reaction provides a simple analytical method for the quantitative determination of boron nitride. With elemental fluorine, the same authors obtained quantitative yields of trifluoroborane and nitrogen.

$$2BN + 3F_2 \rightarrow 2BF_3 + N_2 \tag{VI-4}$$

It is of interest to note that neither reaction (VI-3) or (VI-4) provides any fluorinated borazine derivatives. This observation, however, is in line with a report that $H_3N \cdot BF_3$ decomposes irreversibly to NH_4BF_4 and boron nitride at 125—150° and no intermediate species has been isolated[7]. Boron nitride is not attacked by other mineral acids and, in general, has been found to be very resistant to other kinds of chemical attack. More recent data[8] indicate that hot, concentrated alkali cleaves the boron-nitrogen bond, but water will not attack the nitride. Oxidation of boron nitride in air begins at temperatures above 1200°. The melting point of boron nitride is near 3000°, but dissociation starts in vacuum near 2700°. This high chemical and thermal stability enables one to utilize boron nitride as a crucible material.

[1] STOCK, A., and F. BLIX: Ber. dtsch. chem. Ges. **34,** 3046 (1901).
[2] STOCK, A., and W. HOLLE: ibid. **41,** 2095 (1908).
[3] FRIEDERICH, E., and V. SITTIG: Z. anorg. allg. Chem. **143,** 293 (1925).
[4] WEINTRAUB, G.: U.S. Patent 1,157,271 (1915).
[5] BALMAIN, W. H.: J. prakt. Chem. **32,** 494 (1844).
[6] GLEMSER, O., and H. HAESELER: Z. anorg. allg. Chem. **279,** 141 (1955).
[7] LAUBENGAYER, A. W., and G. F. CONDIKE: J. Amer. chem. Soc. **70,** 2274 (1948).
[8] RENNER, TH.: Z. anorg. allg. Chem. **298,** 22 (1958).

Appendix

^{11}B Nuclear Magnetic Resonance Spectroscopy of Boron-Nitrogen Compounds

A. General Remarks

The element boron has two naturally occurring isotopes; the natural abundance of ^{10}B is 18.3%, that of ^{11}B is 81.7%. Both nuclei have magnetic moments, but normally only the effects of the boron isotope ^{11}B are considered in boron nuclear magnetic resonance studies. The ^{11}B atom has a spin of $I = 3/2$. The first ^{11}B resonance studies were reported by OGG[1] and SHOOLERY[2, 3]; ^{11}B nuclear magnetic resonance spectroscopy has played an important part in the structural determination of boron hydrides[4].

In a qualitative and somewhat simplified approach, the major features of the ^{11}B chemical shifts can be accounted for on the basis of a paramagnetic shift. This shift arises from the difference in electron occupancy of the bonding p_x and p_y orbitals and the partially filled p_z orbital. Neglecting hyperconjugation, the bonding hybrid of boron in trimethylborane, $B(CH_3)_3$, should be sp^2 with a vacant p_z orbital. This characteristic accounts for low field resonance in the ^{11}B nuclear magnetic resonance spectrum. In other molecules of the type BR_3, contributions from electrons of atoms bonded to boron provide partial double bond character between the boron and its substituent. Thus the valence shell of boron approaches an electron octet and the ^{11}B chemical shift moves toward a higher field. At the other end of the spectrum, opposed to trimethylborane, the ^{11}B resonance of the hydroborate anion, $BH_4^{\ominus}$, should be at its highest field. The hydroborate ion presumably exists as a sp^3 bonding hybrid of boron with full tetrahedral symmetry about that atom. This situation is evidenced by high field ^{11}B nuclear magnetic resonance.

If the effect of the paramagnetic shift governs shielding of boron, the ^{11}B chemical shifts of all boron compounds should be expected to fall between the values of trimethylborane at low field and the hydroborate ion at high field, an area which stretches over about 130×10^{-6}. Indeed, all boron compounds which have been studied thus far with ^{11}B nuclear magnetic resonance spectroscopy have their ^{11}B chemical shift values between these two. The only exceptions known are the

[1] OGG, R. A.: J. chem. Physics **22**, 1933 (1954).

[2] SHOOLERY, J. N.: Disc. Faraday Soc. **19**, 215 (1955).

[3] SCHAEFFER, R., J. N. SHOOLERY and R. JONES: J. Amer. chem. Soc. **79**, 4606 (1957).

[4] LIPSCOMB, W. N.: "Boron Hydrides", W. A. Benjamin, Inc., New York—Amsterdam, 1963.

apex atom in B_5H_9 and some triiodoborane complexes[1], which exhibit their ^{11}B resonance at even higher field than the hydroborate ion. Solvent effects upon the chemical shifts of ^{11}B resonance spectra appear to be negligible, except, of course, when chemical reaction occurs between the boron compound and the solvent. Therefore, the ^{11}B chemical shift of a boron compound can be regarded as a characteristic constant for identification purposes.

The diethyletherate of trifluoroborane, $(C_2H_5)_2O \cdot BF_3$, is normally used as the zero reference for ^{11}B chemical shift data, because of the sharp resonance line and the ready availability of the compound. However, in many instances, trimethoxyborane, $B(OCH_3)_3$, has been used. The chemical shifts of these two major zero references for ^{11}B resonance spectroscopy differ by a value of $+18.1 \times 10^{-6}$ for the etherate of trifluoroborane relative to that of trimethoxyborane.

The two major groups of ^{11}B chemical shifts as described above (i.e. sp^2 or sp^3 hybridization) can be characterized further on the basis of inductive effects of the atoms bonded directly to boron, since non-adjacent atoms have little effect on the shielding. On this basis, certain series of ^{11}B chemical shifts can be observed. They are essentially in agreement with the polar bond theory which requires that shifts become more negative as the electronegativity of the substituents increases. This situation is illustrated by the fact that the shielding of a boron nucleus in the trihalogenoborane series decreases in the order $BI_3 > BBr_3 > BCl_3$. Only trifluoroborane exists as a exception. The high shielding of boron in trifluoroborane can readily be explained in terms of the increased ionic character of this compound through partial formation of a B—F double bond, i.e. occupation of p_z orbitals of boron by non-bonding electrons of the fluorine atoms.

Summarizing, there are two main factors which contribute to the shielding of boron nuclei and thus to the actual values of ^{11}B chemical shift data. Of major importance is hybridization of the boron in the direction $sp^2 \to sp^3$ in conformance with a chemical shift to higher field. However, the electronegativity of ligands plays a distinct role in the nuclear shielding of boron atoms. — There are no references available on nitrogen resonance studies of boron-nitrogen compounds.

A comparison of spin-spin coupling constants between boron and hydrogen in a variety of boron compounds illustrates that the coupling constants are smaller if hybridized sp^3 orbitals are used for bonding instead of sp^2 orbitals. Thus a constant $J_{BH} = 81$ cps. is found for the hydroborate anion, whereas in borazine, $(-BH-NH-)_3$, it is 136 cps. Coupling constants between these two values have been measured for those derivatives of the hypothetical BH_3 where one assumes a boron hybridization between sp^2 and sp^3. In general, the widths of ^{11}B resonance lines are considerable due to quadrupole relaxation effects. Consequently, spin-spin coupling between boron and hydrogen beyond atoms

---

[1] LANDESMAN, H., and R. E. WILLIAMS: J. Amer. chem. Soc. **83**, 2663 (1961).

bonded directly to each other has not been observed. In view of the wide lines, the half-band widths of ^{11}B resonances are often cited in conjunction with chemical shift data.

B. ^{11}B Nuclear Magnetic Resonance of Boron-Nitrogen Compounds

When closely related compounds are involved, the value of the ^{11}B chemical shift may be considered roughly proportional to the electron shielding around a boron nucleus. Higher chemical shift values thus indicate greater electronic shielding around the boron. In the case of boron-nitrogen compounds, this condition appears to correspond to increased boron-nitrogen bonding. Some specific applications of ^{11}B nuclear magnetic resonance spectroscopy have been discussed in previous chapters. Hence the following description is limited to some general considerations.

Borazines. As is evident from the chemical shift values of the ^{11}B resonance of borazines, the shielding of a boron nucleus in such molecules is somewhat greater than that in trialkylboranes. This increased shielding can be related to an electron delocalisation: Lone-pair electrons of the nitrogen atoms participate in the bonding between boron and nitrogen, thus providing for resonance structures as discussed previously (Chapter III). Chemical shift values for borazines have been reported in the range from about -32 to -22×10^{-6}.

Aminoboranes. Physical and chemical evidence suggests the existence of double bond character for the boron-nitrogen bond in monomeric aminoboranes. This is evidenced in their ^{11}B resonance spectra through observation of a ^{11}B chemical shift to higher field than one finds with trisubstituted boranes such as trialkylboranes. The ^{11}B chemical shift of organic substituted aminoboranes is found in the -45 to -35×10^{-6} region. Dimerization changes the environment of the boron atom (see Chapter II) and the chemical shift of a dimeric aminoborane is normally found near -10×10^{-6} to 0. Hence ^{11}B chemical shift data can be used for analytical purposes in those cases where partial dimerization takes place: Relative intensities of high and low field ^{11}B resonances permit elaboration of the monomer-dimer ratio. The ^{11}B chemical shift values of trimeric aminoboranes are centered around 0 to $+10 \times 10^{-6}$.

Amine-Boranes. Formation of an amine adduct with a BR_3 entity is accompanied by a change of boron hybridization presumably from sp^2 to sp^3. This denotes partial electron occupation of the boron p_z orbital through donation of the lone-pair electrons from nitrogen. This event is confirmed by the observation of a reduced paramagnetic shift of

Table 1. ^{11}B *Chemical Shift Data*

Compound	Solvent	X 10^{-6}	References
$B(CH_3)_3$	neat	−86.3	7
$B(C_2H_5)_3$	neat	−84.7	6, 7
$(—BH—NCH_3—)_3$	neat	−32.4	7
$(—BH—NH—)_3$	neat	−30.4	7
$(—BCHCH_2—NH—)_3$	neat	−31.8	2
$(—BF—NH—)_3$	benzene	−25.1	5
$(—BF—NCH_3—)_3$	benzene	−24.3	5
$(—BF—NC_3H_7—)_3$	benzene	−23.1	5
$(C_2H_5)_2N—BHC_4H_9$ (i)	neat	$\begin{cases}-47.5\\-38.1\end{cases}$	4
$(CH_3)_2N—BHC_4H_9$ (t)	neat	$\begin{cases}-48.5\\-38.7\end{cases}$	4
$(CH_3)_2N—B(C_4H_9)_2$	neat	−45.3	8
$(CH_3)\,(C_6H_5)N—B(CH_3)\,(C_6H_5)$	neat	−44.3	1
$n\text{-}C_4H_9B[N(CH_3)_2]_2$	neat	−34.2	8
$n\text{-}C_4H_9B[N(C_2H_5)_2]_2$	neat	−39.3	4
$C_6H_5B[N(CH_3)_2]_2$	neat	−32.4	8
$B[N(CH_3)_2]_3$	neat	−27.1	8
$B[N(C_2H_5)_2]_3$	neat	−31.0	7
$[H_3CHN—BHC_4H_9(t)]_2$	neat	$\begin{cases}-8.4\\-0.1\end{cases}$	4, 3
$(—BH_2—NH_2—)_3$	ammonia	+11.3	3
$(—BH_2—NHCH_3—)_3$	methanol	+5.8	3
$(—BCl_2—NH_2—)_3$	diglyme	−6.9	3
$(—BHCl—NHCH_3—)_3$	chloroform	−1.1	7
$B_2H_4[N(CH_3)_2]_2$	neat	+18.6	7
$B_2H_5N(CH_3)_2$	neat	−3.6	7
$(CH_3)_3N \cdot BF_3$	benzene/methanol	+0.5	7
$H_3N \cdot BF_3$	water	+2.1	6
piperidine $\cdot BF_3$	carbon disulfide	+2.3	6
NH_4BF_4	water	+1.8	6
$(CH_3)_3N \cdot BH_3$	benzene	+9.1	6
$(CH_3)_2HN \cdot BH_3$	benzene	+15.1	7
pyridine $\cdot BH_3$	neat	$\begin{cases}+13.3\\+11.5\end{cases}$	6
$NaBH_4$	0.1 n NaOH	+42.9	7
	water	+38.7	6

[1] BAECHLE, H., H. J. BECHER, H. BEYER, W. S. BREY, J. W. DAWSON, M. E. FULLER II and K. NIEDENZU, Inorg. Chem. **2**, 1065 (1963).

[2] FRITZ, P., K. NIEDENZU and J. W. DAWSON: ibid. **3**, 626 (1964).

[3] GAINES, D. F., and R. SCHAEFFER: J. Amer. chem. Soc. **85**, 3592 (1963).

[4] HAWTHORNE, M. F.: ibid. **83**, 2671 (1961).

[5] NIEDENZU, K., H. BEYER and H. JENNE: Chem. Ber. **96**, 2649 (1963).

[6] ONAK, T. P., H. LANDESMAN, R. E. WILLIAMS and I. SHAPIRO: J. physic. Chem. **63**, 1533 (1959).

[7] PHILLIPS, W. D., H. C. MILLER and E. L. MUETTERTIES: J. Amer. chem. Soc. **81**, 4496 (1959).

[8] RUFF, J. K.: J. org. Chemistry **27**, 1020 (1962).

the ¹¹B resonance. The ¹¹B resonance of the amine adducts of tri-fluoroborane are essentially coincident with that of the tetrafluoro-borate anion, $BF_4^{\ominus}$, suggesting that the electronic configuration around the boron atom is similar for both types of compounds. Therefore, it can be concluded that boron in amine-trifluoroboranes is a tetrahedrally bonded sp^3 hybrid. Indeed, X-ray work has confirmed a nearly tetra-hedral configuration of boron in some of these compounds.

Table 1 describes the ¹¹B chemical shifts of some boron-nitrogen compounds relative to the diethyletherate of trifluoroborane. Literature values with zero references other than this one are recalculated by using the proper chemical shift data of the actual reference as compared to trifluoroborane diethyletherate.

Author Index

Abel, E. W., J. D. Edwards, W. Gerrard and M. F. Lappert 38
—, W. Gerrard, M. F. Lappert and R. Shafferman 42, 43
Abraham, M. A., J. H. N. Garland, J. H. Hill and L. F. Larkworthy 34
Adams, M. D. see Schaeffer, G. W. 52, 64
Adams, R. M., and F. D. Poholsky 46
Anderson, E. R. see Schaeffer, G. W. 14, 54, 65, 87, 93, 111
Andrews, E. R. see Zimmer, H. 140
Apple, E. F. and T. Wartik 80
Aries R. S. 72
Aristorkhova, G. I. see Korshak, V. V. 108
Aronovich, P. M. see Mikhailov, B. M. 66, 96
Ashby, E. C. 21
—, and W. E. Foster 14
— see McCusker, P. A. 56
Aubrey, D. W., W. Gerrard and E. F. Mooney 71, 72
—, and M. F. Lappert 97, 119, 125
—, — and M. K. Majumdar 71, 72, 73
—, — and H. Pyszora 72

Baechle, H., H. J. Becher, H. Beyer, W. S. Brey, J. W. Dawson, M. E. Fuller II and K. Niedenzu 49, 157
Balmain, W. H. 2, 153

Banus, J. see Burg, A. B. 57
Barbaras, G. K. see Brown, H. C. 17
Barfield, P. A., M. F. Lappert and J. Lee 49
Bartlett, R. K., H. S. Turner, R. J. Warne, M. A. Young and W. S. McDonald 120
Basile, L. J. see Schaeffer, G. W. 64
Battenberg, E. see Meerwein, H. 29
Bauer, S. H. 88, 150
—, G. R. Finlay and A. W. Laubengayer 27, 29
Beachley jr., O. T. see Laubengayer, A. W. 123
Bean, F. R., and J. R. Johnson 46
Becher, H. J. 32, 49, 50, 51, 57, 72
—, and S. Frick 94, 99, 115, 116, 118
—, W. Nowodny, H. Nöth and W. Meister 80, 84
— see Baechle, H. 49, 157
— see Goubeau, J. 10, 11, 23, 48, 50, 59
Becke-Goehring, M., and H. Krill 56
— and W. Lehr 87
Bekasova, N. I. see Korshak, V. V. 108
Bellamy, L. J., W. Gerrard, M. F. Lappert and R. L. Williams 42
Benson, R. E. see Holmquist, H. E. 47
Berzelius, J. J. 35
Besson, A. 35, 40, 41

Beyer, H., J. W. Dawson, H. Jenne and K. Niedenzu 62, 70, 101
—, J. B. Hynes, H. Jenne and K. Niedenzu 90, 99, 117, 119
—, K. Niedenzu and J. W. Dawson 97, 138, 139, 140
— see Baechle, H. 49, 157
— see Niedenzu, K. 49, 51, 54, 55, 57, 60, 67, 69, 70, 77, 79, 99, 104, 129, 130, 134, 135, 157
— see Nöth, H. 13, 15, 18, 19, 20, 21, 24, 29, 37, 38, 41
Bidinosti, D. R. see Laubengayer, A. W. 99, 104, 107
Birnbaum, E. R. see Ryschkewitsch, G. E. 19
Bissot, T. C., and R. W. Parry 52, 63
Blau, J. A., W. Gerrard and M. F. Lappert 61
Blix, F. see Stock, A. 153
Block, F. see McLaughlin, D. E. 17
Blokhina, A. N. see Mikhailov, B. M. 66, 96, 105, 111
Bloomfield, J. J. see Mulvaney, J. E. 140
Böddeker, K. W. see Shore, S. G. 64
Bohn, U. see Goubeau, J. 42
Bolz, A. see Wiberg, E. 52, 54, 58, 66, 86, 87, 92, 94, 102, 109, 118
Boone, J. L., and G. W. Willcockson 108

Harrelson, D. H. see
Niedenzu, K. 55, 56,
58, 96, 100, 101,
109, 110, 112
Harris, J. J. 124
—, and B. Rudner 104
—, and Ph. D. Thesis
127
— see Rudner, B. 97,
139
—. see Ryschkewitsch,
G. E. 96, 99, 100, 105,
111, 112, 118, 120
Hawkins, R. T.,
W. J. Lennarz and
H. R. Snyder 138
—, and H. R. Snyder
138, 139
Haworth, D. T., and
L. F. Hohnstedt 87,
92, 93, 109, 111
— see Hohnstedt, L. F.
88, 96, 98, 105,
Hawthorne, M. F. 16,
17, 18, 19, 21, 53,
54, 57, 66, 69, 94, 138,
139, 157
—, and E. S. Lewis 19
—, and A. R. Pitochelli
35
— see Lipscomb, W. N.
35
Heal, H. G. 62
Hein, F., and
R. Burckhardt 44
Heller, H. H. see
Meulen, P. A. van der
29, 31
Hennion, G. F. see
McCusker, P. A. 56
Hermann, C. see
Brill, R. 151
Hertwig, K. see
Wiberg, E. 11, 17,
52, 54, 59, 87, 94
Herzberg, G. see
Douglas, A. E. 152
Hesse, G., and H. Witte
47
Heying, T. L., and
H. D. Smith jr. 62
Hickam, C. W. see
Shore, S. G. 64
Hill, J. H. see
Abraham, M. A. 34
Hill, L. see Muszkat,
K. A. 99, 102, 104,
105
Hoard, J. L., S. Geller
and W. N. Cashin 31

Hoard, J. L., S. Geller and
T. B. Owen 31
—, T. B. Owen,
A. Buzzell and
O. N. Salmon 31, 39
— see Coursen, D. L.
114
— see Geller, S. 31
Hoffmann, A. K., and
W. M. Thomas 46
Hoffmann, R. 49, 50,
51, 65, 142, 143
Hohnstedt, L. F., and
D. T. Haworth 88,
96, 98, 105
—, and R. F. Leifield 96
—, and A. M. Pellicciotto
139
— see Haworth, D. T.
87, 92, 93, 109, 111
— see Schaeffer, R. 88,
98, 103, 121
Holle, W. see Stock, A.
40, 148, 153
Hollerer, G. see Nöth, H.
67
Holliday, A. K.,
F. J. Marsden and
A. G. Massey 79, 80
—, and A. G. Massey 80
—, — and F. B. Taylor 80
—, and N. R. Thompson
20
Holmes, R. R. see
Brown, H. C. 10, 33
Holmquist, H. E., and
R. E. Benson 47
Homberg, G. 1
Horeld, G. see Wiberg, E.
59, 96, 104
Horn, H., and
E. S. Gould 42
Horvitz, L. see
Schlesinger, H. I. 54,
87, 94, 118
Hough, W. V., and
G. W. Schaeffer 53,
64
Howarth, M. see
Gerrard, W. 96
Hudson, H. R. see
Gerrard, W. 100,
101, 112, 124
Hückel, W. 151
Hughes, E. W. 23
Hunt, D. W. see
Brown, M. P. 81, 82
Hunter, D. L. see
Steinberg, H. 46

Hynes, J. B. see Beyer,
H. 90, 99, 117, 119

Inatome, M., and
L. P. Kuhn 34
— see Kuhn, L. P. 35
Ito, K., H. Watanabe
and M. Kubo 91,
117, 118
— see Watanabe, H. 90,
115
Iwasaki, K. see
Köster, R. 55, 61,
73, 75, 136

Jacobs, L. E.,
J. R. Platt and
G. W. Schaeffer 90
Jenne, H., and
K. Niedenzu 56, 62,
74, 75, 97
— see Beyer, H. 62, 70,
90, 99, 101, 117, 119
— see Niedenzu, K. 49,
55, 57, 60, 99, 104,
129, 157
Joannis, A. 2, 36, 40,
41, 95
Johannesen, R. B. see
Brown, H. C. 18
Johnson, A. R. 40, 41
Johnson, J. R. see
Bean, F. R. 46
Johnson, L. see Camp-
bell, G. W. 63
Johnson, S. see Brown,
H. C. 27, 29
Johnson, W. H. see
Kilday, M. V. 91
Jones, R. see
Schaeffer, R. 154
Jones, R. G., and
C. R. Kinney 36, 71,
95, 105, 110

Kaczmarczky, A.,
R. D. Dobrott and
W. N. Lipscomb 35
Kampmann, F. W. see
Goubeau, J. 28
Kato, S., M. Wada and
Y. Tsuzuki 135
Kaufman, J. J., and
J. R. Hamann 51,
146, 147
— see Chalvet, O. 90,
114

166 Author Index

Miller, M. A. 33
Miller, N. E.,
  B. L. Chamberland
  and E. L. Muetter-
  ties 26
—, and E. L. Muetterties
  26
Miller, R. R. see
  Eddy, L. B. 88
Mitschelen, H. see
  Goubeau, J. 28, 29
Mixter, W. G. 27, 29
Moews jr., P. C., and
  A. W. Laubengayer 93
— see Lauben-
  gayer, A. W. 91, 122
Mooney, E. F. 117, 118
— see Aubrey, D. W.
  71, 72
— see Burch, J. E. 67,
  69, 96, 97
— see Butcher, I. M.
  116, 117
— see Gerrard, W. 12,
  29, 36, 96, 100, 101,
  105, 110, 112, 124
Moore, R. E. see Urry, G.
  79, 80
Mootz,D. see Luther, H. 23
Morris, R. C. see
  Conklin, G. W. 133
Mountfield, B. A. see
  Gerrard, W. 62
Muetterties, E. L. 25
—, and E. G. Rochow 32
— see Brown, C. A. 29,
  30, 32
— see Miller, N. E. 26
— see Phillips, W. D.
  16, 53, 69, 157
Mulliken, R. S. 9, 27
— see Roothaan, C. C. J.
  90
Mulvaney, J. E.
  J. J. Bloomfield and
  C. S. Marvel 140
Musgrave, O. C. 53
—, and T. O. Park 45
Muszkat, K. A., L. Hill
  and B. Kirson 99,
  102, 104, 105
—, and B. Kirson 99,
  102, 104, 105, 109,
  110

Nakagawa, T. see
  Ohashi, O. 20
— see Totani, T. 49
— see Watanabe, H. 89,
  116

Nakamura, D., H. Wata-
  nabe and M. Kubo
  114
Narisada, M. see
  Watanabe, H. 116
Nazy, J. R. see
  Letsinger, R. L. 43
Nespital, W. 39
— see Ulich, H. 39
Neu, R. 44, 138, 139
Newsom, H. C.,
  W. D. English,
  A. L. McCloskey and
  W. G. Woods 106
—, W. G. Woods and
  A. L. McCloskey
  106, 120
Niedenzu, K. 103, 104,
  128
—, H. Beyer and
  J. W. Dawson 51, 55,
  67, 69, 70, 77, 79,
  129, 130, 134, 135
—, —, — and H. Jenne
  49, 55, 57, 60
—, — and H. Jenne 99,
  104, 157
—, and J. W. Dawson
  49, 50, 57, 58, 59,
  60, 100, 102, 105,
  106, 112
—, — and W. George
  101, 102
—, P. Fritz and
  J. W. Dawson 76,
  134, 135
—, W. George and
  J. W. Dawson 100,
  101
—, D. H. Harrelson and
  J. W. Dawson 55, 96,
  100, 101, 109, 110
—, —, W. George and
  J. W. Dawson 56, 58
—, H. Jenne and
  P. Fritz 128
— and coworkers 75,
  98, 107, 122
— see Baechle, H. 49,
  157
— see Beyer, H. 62, 70,
  90, 97, 99, 101, 119,
  138, 139, 140
— see Fritz, P. 107,
  134, 157
— see Jenne, H. 56, 62,
  74, 75, 97,
— see Wyman, G. M.
  50, 51 56

Nieuwland, J. A. see
  Bowlus, H. 29, 32
— see Sowa, F. J. 33
Nöth, H. 55, 56, 57,
  73, 75, 76, 97, 103,
  129, 134
—, and H. Beyer 13, 15,
  18, 19, 20, 21, 24, 29,
  37, 38, 41,
—, — and H. J. Vetter
  24
—, W. A. Dorochov,
  P. Fritz and F. Pfab
  59, 60, 67, 70
—, and P. Fritz 59, 60,
  67, 68, 69, 71, 81, 84
—, and G. Hollerer 67
—, and S. Lukas 58, 59,
  67, 72
—, and W. Meister 67,
  80, 81, 82, 83, 84
—, and W. Regnet 77,
  78, 79, 129, 130
—, H. Schick and
  W. Meister 83
—, and G. Schmid 69
— see Becher, H. J. 80,
  84
Nordman, C. E. 27
—, and C. R. Peters 22,
  23
—, and C. Reimann 23
— see Parry, R. W. 26
Nowodny, W. see
  Becher, H. J. 80, 84
Nyilas, E., and
  A. H. Soloway 133,
  138, 139

O'Connor, T. E. 148
— see Thomas jr., J.
  148
Ogg, R. A. 154
Ohashi, O., Y. Kurita,
  T. Totani, H. Wata-
  nabe, T. Nakagawa
  and M. Kubo 20
— see Totani, T. 49
Okamato, Y., and
  A. J. Gordon 113
Onak, T. P.,
  H. Landesman,
  R. E. Williams and
  I. Shapiro 157
Onyszchuk, M. see
  Paterson, W. G. 29,
  32, 33
Orgel, L. E. see
  Longuet-Higgins,
  H. C. 128

Subject Index

Bold-face type denotes reference(s) of major import

Table Index